# The CIO's Guide to Risk

# The CIO's Guide to Risk

Jessica Keyes

## CRC Press
Taylor & Francis Group
Boca Raton  London  New York

CRC Press is an imprint of the
Taylor & Francis Group, an **informa** business

AN AUERBACH BOOK

CRC Press
Taylor & Francis Group
6000 Broken Sound Parkway NW, Suite 300
Boca Raton, FL 33487-2742

© 2018 by Taylor & Francis Group, LLC
CRC Press is an imprint of Taylor & Francis Group, an Informa business

No claim to original U.S. Government works

Printed on acid-free paper

International Standard Book Number-13: 978-1-1380-9036-1 (Hardback)

**Library of Congress Cataloging-in-Publication Data**

Names: Keyes, Jessica, 1950- author.
Title: CIO's guide to risk / Jessica Keyes.
Description: Boca Raton, FL : CRC Press, 2017.
Identifiers: LCCN 2017030802| ISBN 9781138090361 (hb : alk. paper) | ISBN 9781315108674 (e)
Subjects: LCSH: Information technology projects--Management. | Risk management.
Classification: LCC HD30.2 .K4647 2017 | DDC 004.068/1--dc23
LC record available at https://lccn.loc.gov/2017030802

Visit the Taylor & Francis Web site at
http://www.taylorandfrancis.com

and the CRC Press Web site at
http://www.crcpress.com

This book is dedicated to my family and friends.

# Contents

# Acknowledgments

I would especially like to thank those who assisted me in putting this book together. As always, my editor, John Wyzalek, was instrumental in getting my project approved and providing great encouragement.

# Author

**Jessica Keyes** is president of New Art Technologies, Inc., a high-technology and management consultancy and development firm started in New York in 1989.

Keyes has given seminars for such prestigious universities as Carnegie Mellon, Boston University, University of Illinois, James Madison University, and San Francisco State University. She is a frequent keynote speaker on the topics of competitive strategy, and productivity and quality. She is former adviser for DataPro, McGraw-Hill's computer research arm, as well as a member of the Sprint Business Council. Keyes is also a founding board of directors member of the New York Software Industry Association. She completed a two-year term on the Mayor of New York City's Small Business Advisory Council. She currently facilitates doctoral and other courses for the University of Phoenix and is a member of the Faculty Council for the College of Information Systems & Technology. She has been the editor for WGL's *Handbook of eBusiness* and CRC Press's *Systems Development Management* and *Information Management*.

Prior to founding New Art, Keyes was managing director of R&D for the New York Stock Exchange and has been an officer with Swiss Bank Co. and Banker's Trust, both in New York City. She holds a

Master of Business Administration from New York University, and a doctorate in management.

A noted columnist and correspondent with over 200 articles published, Keyes is the author of the following books:

*The New Intelligence: AI in Financial Services* (HarperBusiness, 1990)

*The Handbook of Expert Systems in Manufacturing* (McGraw-Hill, 1991)

*Infotrends: The Competitive Use of Information* (McGraw-Hill, 1992)

*The Software Engineering Productivity Handbook* (McGraw-Hill, 1993)

*The Handbook of Multimedia* (McGraw-Hill, 1994)

*The Productivity Paradox* (McGraw-Hill, 1994)

*Technology Trendlines* (Van Nostrand Reinhold, 1995)

*How to be a Successful Internet Consultant* (McGraw-Hill, 1997)

*Webcasting* (McGraw-Hill, 1997)

*Datacasting* (McGraw-Hill, 1997)

*The Handbook of Technology in Financial Services* (Auerbach, 1998)

*The Handbook of Internet Management* (Auerbach, 1999)

*The Handbook of eBusiness* (WGL, 2000)

*The Ultimate Web Developer's Sourcebook* (AMACOM, 2001)

*How to be a Successful Internet Consultant*, 2nd edition (AMACOM, 2002)

*Handbook of Software Engineering* (Auerbach, 2002)

*Handbook of Configuration Management* (Auerbach, 2004)

*Human Web* (New Art Press, 2004)

*Balanced Scorecard for IT* (Auerbach, 2005)

*Knowledge Management, Business Intelligence and Content Management: The IT Practitioner's Guide* (Auerbach, 2006)

*X-Internet* (Auerbach, 2007)

*Project Management* (Auerbach, 2008)

*Marketing IT Products and Services* (CRC Press, 2009)

*Implementing the Project Management Balanced Scorecard* (CRC Press, 2010)

*Social Software Engineering: Development and Collaboration with Social Networking* (CRC Press, 2011)

*Enterprise 2.0: Social Networking Tools to Transform Your Organization* (Auerbach, 2012)

*Bring Your Own Devices (BYOD) Survival Guide* (CRC Press, 2013)

*The CIO's Guide to Oracle Products and Solutions* (CRC Press, 2014)

*Managing IT Performance to Create Business Value* (CRC Press, 2016)

# Introduction

In an age of globalization, widely distributed systems, and rapidly advancing technological change, IT professionals and their managers must understand that risk is ever present. The key to project success is to identify risk and then deal with it.

The *CIO's Guide to Risk* addresses the many faces of risk, whether it be in systems development, adoption of bleeding edge tech, the push for innovation, and even our march toward all things social media. Risk management planning, risk identification, qualitative and quantitative risk analysis, contingency planning, and risk monitoring and control are all addressed on a macro as well as micro level.

To aid with implementing a risk management program, checklists, plans, templates, Appendices A–P, and other related documents are available for download at: https://www.crcpress.com/9781138090361.

# 1

# ANALYZING TECHNOLOGY TRENDS TO EVALUATE RISK

Technology = Change. The faster the pace of technological advancement, the faster the pace of business change. From mainframes in the 1970s; to PCs in the 1980s; to the Internet in the 1990s; to virtual worlds, social media, and flying cars in the 2000s. Structural change such as this requires the organization to be well-aware of the trends powering these massive changes. Organizations must be reactive to these technology trends as well. To gain a better insight on meeting the technology trend challenge, the organization must delve into the intricacies of researching methods for uncovering technology trends, analyzing the trends, selecting the trends most applicable to the organization, understanding infrastructural changes required as a result of implementing the new technologies, developing metrics for success, and managing risk.

## Technology Trend Analysis

Technology trend analysis utilizes information gathering techniques to collect detailed information on those technologies that hold promise for your organization. The breadth and depth of the research will vary from project to project depending on the needs of the organization. The more in depth the research, the better prepared you will be to make timely decisions. In today's fast pace business environment, timing is often the only thing that separates industry leaders from those that lag behind in terms of achieving competitive advantage. Almost fifty years ago, Fred Smith founded FedEx (Federal Express) using just these techniques. He determined that parcel delivery was economically inadequate. Smith determined that there was a need for shippers to have a computer system designed specifically for airfreight that could accommodate time-sensitive shipments such as medicines, computer parts, and electronics.

Bear in mind that the information gathering process is a continual process that involves updating the information gathered as technologies evolve and converge with other technologies. One of the key information gathering activities involves collecting information not only about technology, but about organizational goals and strategies. This information is usually readily available within various planning documents that the organization generates as part of its regular planning cycle. This information is critical in keeping technology aligned with organizational goals.

The goal, at this point, should be to get as much accurate information as possible. Don't be concerned with analyzing the information at this point, which will happen later in the process. The concern at this point is that the information gathered be accurate and as complete as possible.

The following checklist will help to ensure you have all the information you will need to move forward with the process. It is not necessary to do every item listed, but you should look at the listed items to ensure that all the appropriate bases are covered. Be sure to consider the risk that the data you are collecting presents.

Internal information gathering
- Gather internal organizational information, mission statement, strategic goals, tactical goals, and operational goals
- Distribute a survey to key decision makers and align expectations for primary goals with technology trends
- Gather information on marketing and advertising strategies, and identify short- and long-term goals
- Understand branding strategy including the desired perception, message, and tone
- Gather qualitative data on customers and customer service

External information gathering
- Research your industry using traditional research methods, the library, online searches, paid research organization, etc.
- Research industry-specific publications, newsgroups associations, organizations, white papers, etc.

- Identify primary and secondary competition
- Identify main differentiators from competition

Technology information gathering

- Gather quantitative data on existing infrastructure; use logs to analyze traffic patterns and identify trends
- Conduct a technology audit to determine how effective the current infrastructure is at meeting the overall goals of the organization
- Define the extent of infrastructure upgrades required to support the new technology
- Define the primary goal that the new technology will help the organization achieve
- Create scenarios of what the future would look like if the new technology were implemented

At this phase of the information gathering process, you need to identify the specific goals that the organization needs to achieve by implementing the new technology. The following questions may help to solidify those goals:

- What business objective will the technology impact?
- Can the business objective be tied back to a strategic, tactical, or operational goal?
- What impact will the new technology have on the existing infrastructure?
- Does the organization have internal expertise to support the new technology?

These question need to be addressed but do not get sidetracked; you are not yet looking for the answer to any "how-to" questions. For now, you are just trying to identify possibilities for the new technology. It may be helpful to review some of the information collected during the information gathering phase of the process. Look for specific goals that the technology may help the organization achieve. Some possible goals might include:

- Increase market share
- Increase customer satisfaction
- New product or service development
- Compete more effectively

- Decrease demand for customer service
- Create alliances or partnerships
- Streamline supply chain
- Create a scalable infrastructure for growth

The information gathered should allow you to identify a clear linkage to what organizational goals the technology will address and what obstacles will need to be overcome to effectively integrate the technology into the organization's existing infrastructure.

### Conceptualizing Applicability of Emerging Trends

There are many trends that you could spend your day watching and analyzing. The problem is you would not get any work done. Selecting which trends will have the greatest impact on the organization is pretty simple if you have covered all the bases during the information gathering process. You just look at those technology trends that have the biggest impact on achieving organizational goals. Jeff Bezos looked at the trend toward moving things to the Internet and founded Amazon.com as a result.

One of the current trends that many organizations are following is the trend to focus IT resources on the migration to cloud services (e.g., Software as a Service, or SaaS) that allow the organization to work more closely with those that are involved with the organization's value chain. It has been said that the Internet changed everything; and, despite the dot-com bust (between ~2000 and 2002), the innovation that was unleashed during the boom has paved the way for more traditional businesses to leverage existing systems and the Internet to work more closely with the partners, suppliers, and customers.

Cloud and other web technologies will continue to change the way we do business and to provide challenges and risks that need to be overcome. Perhaps the most significant subtrend in this area is the trend toward customer enablement. By putting more functionality in the hands of the firm's customers, organizations are satisfying demand for ubiquitous, anytime access.

To be successful, these products and services need to be flexible without introducing inappropriate levels of risk. Web services are also seen as moving the industry more toward open systems and reusable code thus making system development more efficient.

Conceptualizing the Future

Conceptualizing the future attempts to create a realistic and practical vision of the future for the organization. The outcomes of the conceptualization process are three critical pieces of information that serve as a roadmap to the future for the organization:

1. The first piece of information developed is used as input into the mission statement. The input serves as a way to keep the mission statement current by defining in broad terms the organization's customers, what value is provided to those customers, and the general methods or means employed to create that value. In other words, the purpose and scope of the business are redefined and boundaries are reestablished. In the past, mission statements were relatively static and changes were seldom made. In the new economy, it is necessary for the organization to continually monitor the environment and periodically adjust the mission statement to keep the vision of the organization in alignment with the trends of the external environment. In the past few years we have seen numerous organizations close their doors because they allowed themselves to become obsolete.

2. The second key output of conceptualizing the future is a statement about the organizational culture of the future. Organizational culture is described as those behaviors that distinguish or set the organization apart from other organizations in the industry. The organizational culture will need to evolve from what it is today to what it needs to be in the future as described in the conceptualization process.

3. The third and final element of the conceptualization process is a statement describing the shared beliefs or underlying values of the organization and those that are employed by the organization. These values should be clearly articulated so as to guide the behavior and decision making that takes place in the organization on a daily basis.

Conceptualizing the future helps leaders assess internal and external strengths, weaknesses, opportunities, and threats (SWOT) and to create buy-in of the vision at all levels of the organization.

**Effect on Infrastructure**

In order to identify elements of the infrastructure that may be affected by new technologies, we need to agree on a definition of the term *infrastructure*. For purposes of this chapter, let us assume that infrastructure refers to those strategic information assets that allow the organization to achieve its mission.

The elements that make up the strategic information assets include:

- Data
- Business processes and applications that transform the data into meaningful information
- Technologies that deliver information to those that need it

By focusing on each of these elements, there is a unique opportunity to leverage and create synergies across the enterprise.

*Data*

The first element on the infrastructure that we want to examine is data. The term *data* is often used interchangeably with *information*. For our purposes, we want to make a clear distinction. Think of data as the raw material fed into our business processes and applications, and transferred into information or business intelligence. Just as with any manufacturing process, the better the raw material going in, the better the end product, in this case information.

Data quality is usually measured in terms of 11 factors. To be useful in terms of business intelligence, a level of quality needs to be defined in each of the 11 areas. The level of quality you seek to achieve for the data should be consistent with the level of quality you seek to achieve in your information. Quality data comes at a cost. Just as you would expect to pay more for a quality automobile, you should be prepared to pay for quality data.

- Data is said to be relevant if it applies to the decision being made. Relevancy is measured in degrees and is best determined by knowledgeable business staff.
- Data is deemed correct if it accurately identifies reality. For some situations, correctness must be absolute, but for others, approximations are adequate in terms of correctness.

- Accuracy is defined as a percentage and is only meaningful when applied to data that has a level of tolerance for error. If there is no tolerance for error, as in the case of an e-mail address, then the accuracy must be 100%.
- Precision can be thought of as a measurement of accuracy for numeric data.
- Completeness means that all the relevant data is available.
- Data needs to be available when needed and it must be a current reflection of reality to be useful.
- Usability refers to the speed and ease that users are able to make use of the data.
- Closely associated with usability is the concept of conforming to the expectations of the user. If you receive a request to identify last year's top 10 customers, you may have several questions that need to be answered before you can proceed. Questions include: Does last year mean the preceding 12 months, or is it the last calendar year or fiscal year? Is the criterion for determining "top customer" total sales volume, net sales volume, profitability, or is it based on quantities? Or, is it a combination of several pieces of data? Each of these questions may have a correct answer for a given department or user. The hard part is getting everyone on the same page in terms of expectation.
- If data is available to those who need access, when and where they need it, and is presented in a usable format, then it is accessible. In the era of Big Data, this is easier said than done.
- *Consistency* is a term used to describe how well all of the data elements that contribute to an item of information work together to provide information that is of value.
- Cost is not really a quality factor; however, it is an important consideration in determining the resources needed to obtain and maintain quality data. It consists of both tangible costs associated with collecting and storing the data (hardware/ software or human resources) and intangible costs (bad decisions being made because quality data is not available). Often the intangible costs are greater than the tangible.

Every piece of data that an organization keeps has an associated quality level. Information technology (IT) professionals need to ensure that quality is not compromised by the introduction of new technologies into the infrastructure.

### Transforming Data

When we talk about technologies that transform data into meaningful information, we are really talking about four separate structures that need to be in place and working together to accomplish this. The four structures are business, information, application, and technology.

- A business structure defines the mission, strategy, business process models and functions, products, and services offered by the organization. This structure keeps the various units within the organization pointed in the same direction and working to achieve a common goal.
- The information structure is a blueprint of how data flows throughout the enterprise. It defines what data is needed by whom, when it is needed, and in what format it is needed in order to accomplish the mission of the organization.
- The application structure focuses on the applications required to support the mission and information needs of the organization. It addresses the common business components and services that can be leveraged to create synergy within the organization.
- A technology structure defines the software, hardware, and network standards needed to support those applications defined in the application structure.

Each of these architectures are tightly coupled and any change introduced into one or more of the structures must be looked at in terms of the impact it will have on the other structures.

### Delivering Information

Managing the telecommunication function involves ensuring that the design is capable of delivering business intelligence to those who need it when they need it and in a format that is usable.

Today's applications and technologies demand ever-increasing amounts of bandwidth and throughput. Rules need to be in place to ensure that connectivity and interoperability is maintained.

To stay current, you need to continually gather information and analyze that information in terms of the organizations strategic goals.

The rate of change in the telecommunication industry has accelerated in recent years to meet the demands of the digital society in which we live. Data traffic now surpasses that of voice traffic on telecommunication networks. Service providers are working hard to provide new technologies to meet what seems to be a never-ending demand for greater speeds, throughput, and anytime anywhere access. New developments in these technologies can create an instant advantage for those organizations that have successfully prepared for their adoption by gathering information and analyzing trends before trying to integrate them into the organization.

One trend that you may want to follow is the increasing usage of wireless technologies. This un-tethered approach to information sharing has been well received by business and IT professionals alike. Mobility has become important in business collaboration and information sharing, and IT professionals need to be prepared to allocate resources to investigating and deploying new wireless devices as they become available.

**Metrics for Measuring Success**

Once you have linked the new technology to organizational goals, you need to determine what criteria you will use to measure success. If you do not define what success is, then you will not be able to determine when you have achieved it.

Oftentimes there are already metrics in place to measure success for existing organizational goals. In this case, the success criteria can be stated in terms of an improvement in the existing measurement. For example, let's say the goal of the new technology is to reduce demand for customer service. Assume that a organization currently tracks calls per day and average time to resolve the customer's issue. In this case, success for the new technology could be defined as a reduction in either one or both of the statistics.

If you are having a hard time identifying success criteria, take your list of goals and sort them into strategic, tactical, and operational goals. For operational goals, you can talk to operational staff to determine how they measure success in their jobs. For tactical goals, you may need to talk to midlevel managers to have them identify what they use for determining if the units they lead are productive. Senior-level executives will be able help you identify measurable ways to determine whether strategic goals have been achieved.

Once you have identified the success criteria, you will need to determine what the baseline for the measurement will be. Determining success criteria and baseline data are critical should the technology be adopted.

### Risk Assessment

Given the poor track record of IT projects, it has become critical for IT managers to perform risk assessments on proposed projects. Risk assessment consists of identifying risks, evaluating the risk, and taking action to manage the risks.

In basic terms, risk is a measurement of the potential of an IT project to fail. IT project failures are usually related to the fluctuation of assumptions that are made during the planning process of a project. Since we know that these assumptions are likely to change, we should consider using a range of values for our assumptions rather that a single value. For example, rather than saying the new backup technology will reduce the time it takes to do nightly backups by 2 hours, you should assign a range, say, from 1.5 to 2.5 hours. Determining which assumptions are likely to fluctuate and what the range of the fluctuation will be can be accomplished by

- Seeking expert advice
- Holding group brainstorming sessions
- Performing assumption analysis

*Expert advice* relies on the experience of others to determine which assumptions are likely to vary and by how much. This is the fastest and easiest method, but care should be taken to ensure that the expert giving the advice is really an expert.

*Group brainstorming sessions* are useful when no experts are available. This is usually the case when implementing new technologies. The group first identifies the assumptions most likely to fluctuate and then determines what the range of fluctuation is most likely to be. This can be a time-consuming process and may require several sessions.

*Assumption analysis* involves the detailed questioning of each assumption to determine under what circumstances the value would change and by how much.

### Risk Evaluation

Risk evaluation is an iterative process that spans the full system development life cycle. Since it needs to be done often, you should have in place processes that allow the evaluation to happen in the most efficient manner possible. A framework that works well for many organizations is to classify risks according to the following:

- Impact on the organization
- Development effort
- Maturity of the technology
- Organizational maturity

Within each category, risks can be classified as high, medium, or low.

Examples of risk factors you should consider for each of the categories are given next:

- Impact on the organization
  - Complexity of the business processes involved
  - Number of organizational units impacted
  - Extent of business rule changes
- Development effort
  - Total cost of project
  - Duration from start to finish
  - Number of project team members
- Maturity of technology
  - Emerging
  - New
  - Proven
- Organizational maturity
  - Organizational track record
  - Stability of the organization
  - Ad hoc processes

The thresholds used to determine if a criterion is a high, medium, or low risk are going to be specific to your organization.

*Risk Management Strategy*

Once risks have been identified and evaluated, you can begin to develop a strategy to manage the risks. Strategies for managing risk include:

- Risk avoidance
- Risk transfer
- Risk assumption

*Risk avoidance* is the elimination of the possibility of loss by not engaging in the activity that produces the risk. In making the decision to avoid a risk, you must weigh the potential value of the activity against the potential loss. An example would be to choose to use a familiar technology rather than a new or emerging one, even though the potential for competitive advantage by using the new technology is greater.

An effective risk management strategy will usually involve various levels of *risk transfer*. The decision to transfer risk if often driven by pricing or the desire to reduce fluctuations in cost or uncertainty. Risk transfer strategies allow an organization to let someone else assume responsibility for risks associated with a particular activity. Outsourcing is an example of risk transfer. As the business environment continues to change and evolve with greater speed, it becomes critical for organizations to partner with industry experts that know the issues of the day, can anticipate future issues, and are better equipped to handle the risks associated with particular technologies.

In general terms, *risk assumption* is knowing a risk exists and choosing not to take any action to avoid or transfer the risk. In essence, you are voluntarily exposing the organization to the risk. It is common to assume low-risk activities. Think of it as the deductible on you car insurance policy. You are assuming the first $500 or so of exposure to the risk of getting in an accident.

### Conclusion

Information gathering is a continual process that prepares the organization to anticipate and analyze technology trends, and then embrace appropriate technologies at an appropriate time. Conceptualizing the applicability of emerging technologies allows the organization to understand and plan for the infrastructural changes that need to be made in order for the organization to survive in the future. It is equally important to identify, evaluate, and manage risks associated with these new technologies.

# 2

# INFORMATION TECHNOLOGY PROJECT RISK

We will examine the concept of project risk, with careful attention paid to the mitigation of risks. We will examine the different varieties of risks (e.g., business, environment, product, employee) learn how to apply probability to each risk, understand the impact of each risk, and ultimately learn how to devise a contingency plan for each risk.

### The Proactive Risk Strategy

Project risk management addresses the following questions:

- Are we losing sight of goals and objectives as the project moves forward?
- Are we ensuring that the results of the project will improve the organization's ability to complete its mission? The result should be an improvement over the previous process.
- Are we ensuring sufficient funds are available, including funds to address risks?
- Are we tracking implementation to ensure "quicker/better/cheaper" objectives are being met?
- Are we applying appropriate risk management principles throughout the project?
- Are we taking corrective action to prevent or fix problems, rather than simply allocating more money and time into them?
- Have changes in the environment, such as new IT systems or leadership, created new risks that need to be managed?

A proactive risk strategy should always be adopted, as shown in Figure 2.1. It is better to plan for possible risk then have to react to it in a crisis.

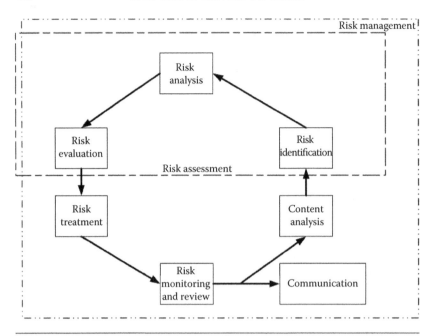

**Figure 2.1**   Risk management feedback loop.

Sound risk assessment and risk management planning through-out project implementation can have a big payoff. The earlier a risk is identified and dealt with, the less likely it is to negatively affect project outcomes. Risks are both more probable and more easily addressed early in a project. By contrast, risks can be more difficult to deal with and more likely to have significant negative impact if they occur later in a project. As explained later, risk probability is simply the likeli-hood that a risk event will occur. Conversely, risk impact is the result of the probability of the risk event occurring plus the consequences of the risk event. Impact, in laymen's terms, is telling you how much the realized risk is likely to hurt.

The propensity (or probability) of project risk depends on the proj-ect's life cycle, which includes five phases: initiating, planning, exe-cuting, controlling, and closing. Although problems can occur at any time during a project's life cycle, problems have a greater chance of occurring earlier due to unknown factors.

The opposite can be said for risk impact. At the beginning of the project, the impact of a problem, assuming it is identified as a risk, is likely to be less severe than it is later in the project life cycle. This is in part because at this early stage there is much more flexibility in making changes and dealing with the risk, assuming it is recognized as a risk. Additionally, if the risk cannot be prevented or mitigated, the resources invested—and potentially lost—at the earlier stages are significantly lower than later in the project. Conversely, as the project moves into the later phases, the consequences become much more serious. This is attributed to the fact that as time passes, there is less flexibility in dealing with problems, significant resources have likely been already spent, and more resources may be needed to resolve the problem.

### Risk Management

The first thing that needs to be done is to identify risks. One method is to create a risk item checklist. A typical project plan might list the following risks:

1. Customer will change or modify requirements
2. Lack of sophistication of end-users
3. Delivery deadline will be tightened
4. End users resist system
5. Server may not be able to handle larger number of users simultaneously
6. Technology will not meet expectations
7. Larger number of users than planned
8. Lack of training of end-users
9. Inexperienced project team
10. System (security and firewall) will be hacked

The optimum approach is to use a framework for identifying software project risks. The following questions are ordered by their relative importance to the ultimate success of a project:

1. Have top software and customer managers formally committed to support the project?
2. Are end-users enthusiastically committed to the project and the system/product to be built?
3. Are requirements fully understood by the software engineering team and their customers?

4. Have customers been fully involved in the definition of requirements?
5. Do end-users have realistic expectations?
6. Is the project scope stable?
7. Does the software engineering team have the right mix of skills?
8. Are project requirements stable?
9. Does the project team have experience with the technology to be implemented?
10. Is the number of people on the project team adequate to do the job?
11. Do all customer/user constituencies agree on the importance of the project and on the requirements for the system/product to be built?

Based on the information uncovered from this questionnaire we can begin to categorize risks. Software risks generally include project risks, technical risks, and business risks.

*Project risks* can include budgetary, staffing, scheduling, customer, requirement, and resource problems. Risks are different for each project, and risks change as a project progresses. Project-specific risks could include, for example, the following:

- Lack of staff buy-in
- Loss of key employees
- Questionable vendor availability and skills
- Insufficient time
- Inadequate project budgets
- Funding cuts
- Cost overruns

*Technical risks* can include design, implementation, interface, ambiguity, technical obsolescence, and leading-edge problems. An example of this is the development of a project around a leading-edge technology that has not yet been proven.

*Business risks* includes building a product or system no one wants (market risk), losing support of senior management (management risk), building a product that no longer fits into the strategic plan (strategic risk), losing budgetary support (budget risks), and building a product that the sales staff does not know how to sell.

A method of risk analysis that requires modularizing the project into measurable parts can be used. Risk can then be calculated as follows:

Exposure factor (EF) = Percentage of asset loss caused by identified threat

Single loss expectancy (SLE) = Asset value × Exposure factor

Annualized rate of occurrence (ARO) = Estimated frequency a threat will occur within a year and is characterized on an annual basis. A threat occurring 10 times a year has an ARO of 10

Annualized loss expectancy (ALE) = Single loss expectancy × Annualized rate of occurrence

Safeguard cost-benefit analysis = (ALE before implementing safeguard) − (ALE after implementing safeguard) − (Annual cost of safeguard) = Value of safeguard to the company

Risks can also be categorized as known, predictable, or unpredictable. Known risks are those that can be uncovered upon careful review of the project plan and the environment in which the project is being developed (e.g., lack of development tools, unrealistic delivery date, or lack of knowledge in the problem domain). Predictable risks can be extrapolated from past experience. For example, your past experience with the end-users has not been good, so it is reasonable to assume that the current project will suffer from the same problem. Unpredictable risks are hard, if not impossible, to identify in advance. For example, no one could have predicted the events of September 11, but this one event affected computers worldwide.

Once risks have been identified, most managers project these risks in two dimensions: likelihood and consequences. As shown in Table 2.1, a risk table is a simple tool for risk projection. First, based on the risk item checklist, list all risks in the first column of the table.

**Table 2.1**  Typical Risk Table

| RISKS | CATEGORY[a] | PROBABILITY | IMPACT[b] |
|-------|-------------|-------------|-----------|
| Risk 1 | PS | 70% | 2 |
| Risk 2 | CU | 60% | 3 |

[a] Category abbreviations: BU, business impact risk; CU, customer characteristics risk; PS, process definition risk; ST, staff size and experience risk; TE, technology risk.
[b] Impact values: 1, catastrophic; 2, critical; 3, marginal; 4, negligible.

**Table 2.2**   Criteria for Determining Likelihood of Occurrence

| LIKELIHOOD: WHAT IS THE PROBABILITY THAT THE SITUATION OR CIRCUMSTANCE WILL HAPPEN? | |
|---|---|
| 5 (Very high) | Very likely to occur. Project's process cannot prevent this event, no alternate approaches or processes are available. Requires immediate management attention. |
| 4 (High) | Highly likely to occur. Project's process cannot prevent this event, but a different approach or process might. Requires management's attention. |
| 3 (Moderate) | Likely to occur. Project's process may prevent this event, but additional actions will be required. |
| 2 (Low) | Not Likely to occur. Project's process is usually sufficient to prevent this type of event. |
| 1 (Very low) | Very unlikely. Project's process is sufficient to prevent this event. |

Then in the following columns, fill in each risk's category, probability of occurrence, and assessed impact. Afterward, sort the table by probability and then by impact, study it, and define a cutoff line (i.e., the line demarking the threshold of acceptable risk).

Table 2.2 describes the generic criteria used for assessing likelihood that a risk will occur. All risks above the designated cutoff line must be managed and discussed. Factors influencing their probability and impact should be specified.

A risk mitigation, monitoring, and management plan (RMMM) is the tool to help avoid risks. Causes to the risks must be identified and mitigated. Risk monitoring activities take place as the project proceeds and should be planned early. Table 2.3 describes typical criteria that can be used for determining consequences of each risk.

**Sample Risk Plan**

An excerpt of a typical RMMM plan is presented next.

## 1.1 SCOPE AND INTENT OF RMMM ACTIVITIES

This project will be uploaded to a server and this server will be exposed to the outside world, so we need to develop security protection. We will need to configure a firewall and restrict access to only "authorized users" through the linked Faculty database. We will have to know how to deal with load balance if the amount of visits to the site is very large at one time.

**Table 2.3**   Criteria for Determining Consequences

| | 1 (VERY LOW) | 2 (LOW) | 3 (MODERATE) | 4 (HIGH) | 5 (VERY HIGH) |
|---|---|---|---|---|---|
| Technical | Minimal or no impact to mission or technical success/exit criteria or margins. Same approach retained. | Minor impact to mission or technical success/exit criteria, but can handle within established margins. Same approach retained. | Moderate impact to mission or technical success/exit criteria, but can handle within established margins. Workarounds available. | Major impact to mission or technical success criteria, but can still meet minimum mission success/exit criteria, threatens established margins. Workarounds available. | Major impact to mission or technical success criteria, cannot meet minimum mission or technical success/exit criteria. No alternatives exist. |
| Schedule | Minimal or no schedule impact, but can handle within schedule reserve; no impact to critical path. | Minor schedule impact, but can handle within schedule reserve; no impact to critical path. | Impact to critical path, but can handle within schedule reserve; no impact to milestones. | Significant impact to critical path, and cannot meet established lower level milestone. | Major impact to critical path and cannot meet major milestone. |
| Cost | Minimal or no cost impact or increase over that allocated, and can be handled within available reserves. | Minor cost impact, but can be handled within available reserves. | Causes cost impact and use of allocated reserves. | Causes cost impact, may exceed allocated reserves, and may require resources from another source. | Causes major cost impact and requires additional budget resources from another source. |

We will need to know how to maintain the database in order to make it more efficient, what type of database we should use, and who should have the responsibility to maintain it and who should be the administrator. Proper training of the aforementioned personnel is very important so that the database and the system contain accurate information.

## 1.2  RISK MANAGEMENT ORGANIZATIONAL ROLE

The software project manager must maintain track of the efforts and schedules of the team. They must anticipate any "unwelcome" events that may occur during the development or maintenance stages and establish plans to avoid these events or minimize their consequences.

It is the responsibility of everyone on the project team with the regular input of the customer to assess potential risks throughout the project. Communication among everyone involved is very important to the success of the project. In this way, it is possible to mitigate and eliminate possible risks before they occur. This is known as a proactive approach or strategy for risk management.

## 1.3  RISK DESCRIPTION

This section describes the risks that may occur during this project.

### 1.3.1  Description of Possible Risks

Business Impact Risk (BU)

> This risk would entail that the software produced does not meet the needs of the client who requested the product. It would also have a business impact if the product no longer fits into the overall business strategy for the company.

Customer Characteristics Risks (CU)

> This risk is the customer's lack of involvement in the project and their nonavailability to meet with the developers in a timely manner. Also the customer's sophistication as to the product being developed and ability to use it is part of this risk.

Development Risks (DE)

> Risks associated with the availability and quality of the tools to be used to build the product. The equipment

and software provided by the client on which to run the product must be compatible to the software project being developed.

Process Definition Risks (PS)

Does the software being developed meet the requirements as originally defined by the developer and client? Did the development team follow the correct design throughout the project? The above are examples of process risks.

Product Size (PR)

The product size risk involves the overall size of the software being built or modified. Risks involved would include the customer not providing the proper size of the product to be developed, and if the software development team misjudges the size or scope of the project. The latter problem could create a product that is too small (rarely) or too large for the client and could result in a loss of money to the development team because the cost of developing a larger product cannot be recouped from the client.

Staff Size and Experience Risk (ST)

This would include appropriate and knowledgeable programmers to code the product as well as the cooperation of the entire software project team. It would also mean that the team has enough team members who are competent and able to complete the project.

Technology Risk (TE)

Technology risk could occur if the product being developed is obsolete by the time it is ready to be sold. The opposite affect could also be a factor: if the product is so "new" that the end-users would have problems using the system and resisting the changes made. A "new" technological product could also be so new that there may be problems using it. It would also include the complexity of the design of the system being developed.

## 1.4  RISK TABLE

The risk table provides a simple technique to view and analyze the risks associated with the project. The risks were listed and then categorized using the description of risks listed in Section 1.3.1. The probability of each risk was then estimated and its impact on the development process was then assessed. A key to the impact values and categories appear at the end of the table.

| RISKS | CATEGORY | PROBABILITY | IMPACT |
|---|---|---|---|
| Customer will change or modify requirements | PS | 70% | 2 |
| Lack of sophistication of end-users | CU | 60% | 3 |
| Users will not attend training | CU | 50% | 2 |
| Delivery deadline will be tightened | BU | 50% | 2 |
| End-users resist system | BU | 40% | 3 |
| Server may not be able to handle larger number of users simultaneously | PS | 30% | 1 |
| Technology will not meet expectations | TE | 30% | 1 |
| Larger number of users than planned | PS | 30% | 3 |
| Lack of training of end-users | CU | 30% | 3 |
| Inexperienced project team | ST | 20% | 2 |
| System (security and firewall) will be hacked | BU | 15% | 2 |

Impact values: 1, catastrophic; 2, critical; 3, marginal; 4, negligible.
Category abbreviations: BU, business impact risk; CU, customer characteristics risk; PS, process definition risk; ST, staff size and experience risk; TE, technology risk.

## 1.5 RMMM STRATEGY

Each risk or group of risks should have a corresponding strategy associated with it. The Risk Mitigation, Monitoring and Management (RMMM) strategy discusses how risks will be monitored and dealt with. Risk plans (i.e., contingency plans) are usually created in tandem with end-users and managers. An excerpt of a RMMM strategy follows.

### 1.5.1 Project Risk RMMM Strategy

The area of design and development that contributes the largest percentage to the overall project cost is the database subsystem. Our estimate for this portion does provide a small degree of buffer for unexpected difficulties (as do all estimates). This effort will be closely monitored and coordinated with the customer to ensure that any impact, either positive or negative, is quickly identified. Schedules and personnel resources will be adjusted accordingly to minimize the effect, or maximize the advantage as appropriate.

Schedule and milestone progress will be monitored as part of the routine project management with appropriate emphasis on meeting target dates. Adjustments to parallel efforts will be made as appropriate should the need arise. Personnel turnover will be managed through use of internal personnel matrix capacity. Our organization has a large software engineering base with sufficient numbers to support our potential demand.

### 1.5.2 Technical Risk RMMM Strategy

We are planning for two senior software engineers to be assigned to this project, both of whom have significant experience in designing and developing web-based applications. The project progress will be monitored as part of the routine project management with appropriate emphasis on meeting target dates, and adjusted as appropriate.

Prior to implementing any core operating software upgrades, full parallel testing will be conducted to ensure compatibility with the system as developed. The application will be developed using only public Application Programming Interfaces (APIs), and no "hidden" hooks. While this doesn't guarantee compatibility, it should minimize any potential conflicts. Any problems identified will be quantified using cost-benefit and trade-off analysis; then coordinated with the customer prior to implementation.

The database subsystem is expected to be the most complex portion of the application, however it is still a relatively routine implementation. Efforts to minimize potential problems include the abstraction of the interface from the implementation of the database code to allow changing the underlying database with minimal impact. Additionally, only industry standard SQL calls will be used, avoiding all proprietary extensions available.

### 1.5.3 Business Risk RMMM Strategy

The first business risk, lower than expected success, is beyond the control of the development team. Our only potential impact is to use the current state-of-the-art tools to ensure performance, in particular database access, meets user expectations; and graphics are designed using industry-standard look-and-feel styles.

Likewise, the second business risk, loss of senior management support, is really beyond the direct control of the development team. However, to help manage this risk, we will strive to impart a positive attitude during meetings with the customer, as well as present very professional work products throughout the development period.

One of the most popular of tools is the risk information sheet, an example of which appears in Table 2.4.

### Risk Avoidance

Risk avoidance, which we introduced in the last chapter, can be accomplished by evaluating the critical success factors (CSFs) of a business or business line. Managers are intimately aware of their missions and goals, but they do not necessarily define the processes required to achieve these goals. In other words, how are you going to

**Table 2.4**   Sample Risk Information Sheet

| RISK INFORMATION SHEET |
| --- |

Risk id: PO2-4-32
Date: March 4, 2017
Probability: 80%
Impact: High
Description:
Over 70% of the software components scheduled for reuse will be integrated into the application. The remaining functionality will have to be custom developed.
Refinement/context:
1. Certain reusable components were developed by a third party with no knowledge of internal design standards.
2. Certain reusable components have been implemented in a language that is not supported on the target environment.
Mitigation/monitoring:
1. Contact third party to determine conformance to design standards.
2. Check to see if language support can be acquired.
Management/contingency plan/trigger:
Develop a revised schedule assuming that 18 additional components will have to be built.
Trigger: Mitigation steps unproductive as of March 30, 2017
Current status:
In process
Originator: Jane Manager

get there? In these instances, technologists must depart from their traditional venue of top-down methodologies and employ a bottom-up approach. They must work with the business units to discover the goal and work their way up through the policies, procedures, and technologies that will be necessary to arrive at that particular goal. For example, the goal of a fictitious business line is to be able to cut the production/distribution cycle by a factor of 10, providing a customized product at no greater cost than that of the generic product in the past. To achieve this goal, the technology group needs to get the business managers to walk through the critical processes that need to be invented or changed. It is only at this point that any technology solutions are introduced.

One technique, called process quality management or PQM, uses the CSF concept. This approach combines an array of methodologies to solve a persistent problem: How do you get a group to agree on goals and ultimately deliver a complex project efficiently, productively, and with a minimum of risk?

PQM is initiated by gathering, preferably off site, a team of essential staff. The team's components should represent all facets of the project. Obviously, all teams have leaders and PQM teams are no different. The team leader chosen must have a skill mix closely attuned to the projected outcome of the project. For example, in a PQM team where the assigned goal is to improve plan productivity, the best team leader just might be an expert in process control, albeit the eventual solution might be in the form of enhanced automation.

Assembled at an off-site location, the first task of the team is to develop, in written form, specifically what the team's mission is. With such open-ended goals as "Determine the best method of employing technology for competitive advantage," the determination of the actual mission statement is an arduous task, best tackled by segmenting this rather vague goal into more concrete subgoals.

In a quick brainstorming session, the team lists the factors that might inhibit the mission from being accomplished. This serves to develop a series of one-word descriptions. Given the 10-minute time frame, the goal is to get as many of these inhibitors as possible without discussion and without criticism.

It is at this point that the team turns to identifying the CSFs, which are the specific tasks that the team must perform to accomplish its mission. It is vitally important that the entire team reach a consensus on the CSFs.

The next step of PQM is to make a list of all tasks necessary in accomplishing the CSF. The description of each of these tasks, called business processes, should be declarative. Start each with an action word such as study, measure, reduce, negotiate, or eliminate.

Table 2.5 and Figure 2.2 show the resulting project chart and priority graph, respectively, that diagram this PQM technique. The team's

**Table 2.5**   CSF Project Chart

| # | BUSINESS PROCESS | 1 | 2 | 3 | 4 | 5 | 6 | COUNT | QUALITY |
|---|---|---|---|---|---|---|---|---|---|
| | | \<CRITICAL SUCCESS FACTORS\> | | | | | | | |
| P1 | Measure delivery performance by suppliers | x | X | | | | | 2 | B |
| P2 | Recognize/reward workers | | | | | x | x | 2 | D |
| P3 | Negotiate with suppliers | x | X | x | | | | 3 | B |
| P4 | Reduce number of parts | x | X | x | x | | | 4 | D |
| P5 | Train supervisors | | | | | x | x | 2 | C |
| P6 | Redesign production line | x | | x | x | | | 3 | A |
| P7 | Move parts inventory | x | | | | | | 1 | E |
| P8 | Eliminate excessive inventory buildups | x | X | | | | | 2 | C |
| P9 | Select suppliers | x | X | | | | | 2 | B |
| P10 | Measure | | | | x | x | x | 3 | E |
| P11 | Eliminate defective parts | | X | x | x | | | 3 | D |

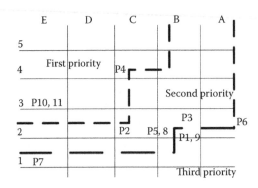

**Figure 2.2**   CSF priority graph.

mission, in this example, is to introduce just-in-time (JIT) inventory control, a manufacturing technique that fosters greater efficiency by promoting stocking inventory only to the level of need. The team, in this example, identified 6 CSFs and 11 business processes labeled P1 through P11.

The project chart is filled out by first ranking the business processes by importance to the project's success. This is done by comparing each business process to the set of critical success factors. A check is made under each critical success factor that relates significantly to the business process. This procedure is followed until each of the business processes have been analyzed in the same way.

The final column of the project chart permits the team to rank each business process relative to current performance, using a scale of A = excellent, to D = bad, and E = not currently performed.

The priority graph, when completed, will steer the mission to a successful, and prioritized, conclusion. The two axes to this graph are quality, using the A through E grading scale, and priority, represented by the number of checks noting each business process received. These can be lifted easily from the project chart for the quality and count columns, respectively.

The final task as a team is to decide how to divide the priority graph into different zones representing first priority, second priority, and so on. In this example, the team has chosen as a first priority all business processes, such as "negotiate with suppliers" and "reduce number of parts," that are ranked from a quality of fair degrading to a quality of not currently performed and having a ranking of three or greater. Most groups employing this technique will assign priorities in a similar manner.

Determining the right project to pursue is one factor in the push for competitive technology. It is equally as important to be able to "do the project right," which can greatly reduce risk.

### Introduction to Quantitative Risk Analysis

Many methods and tools are available for quantitatively combining and assessing risks. The selected method will involve a trade-off between sophistication of the analysis and its ease of use. There are at least five criteria to help select a suitable quantitative risk technique:

1. The methodology should be able to include the explicit knowledge of the project team members about the site, design, political conditions, and project approach.
2. The methodology should allow quick response to changing market factors, price levels, and contractual risk allocation.
3. The methodology should help determine project cost and schedule contingency.
4. The methodology should help foster clear communication among the project team members, and between the team and higher management about project uncertainties and their impacts.
5. The methodology should be easy to use and understand.

Three basic risk analyses can be conducted during a project risk analysis: technical performance analysis (will the project work?), schedule risk analysis (when will the project be completed?), and cost risk analysis (what will the project cost?). Technical performance risk analysis can provide important insights into technology-driven cost and schedule growth for projects that incorporate new and unproven technology. Reliability analysis, failure modes and effects analysis (FMEA), and fault tree analysis are just a few of the technical performance analysis methods commonly used. However, this discussion of quantitative risk analysis will concentrate on cost and schedule risk analysis only.

At a computational level, there are two considerations about quantitative risk analysis methods. First, for a given method, what input data are required to perform the risk analysis? Second, what kinds of data, outputs, and insights does the method provide to the user?

The most stringent methods are those that require as inputs probability distributions for the various performance, schedule, and costs risks. Risk variables are differentiated based on whether they can take on any value in a range (continuous variables) or whether they can assume only certain distinct values (discrete variables). Whether a risk variable is discrete or continuous, two other considerations are important in defining an input probability: its central tendency and its range or dispersion. An input variable's mean and mode are alternative measures of central tendency; the mode is the most likely value across the variable's range. The mean is the value when the variable has a 50 percent chance of taking on a value that is greater and a 50 percent chance of taking a value that is lower.

The other key consideration when defining an input variable is its range or dispersion. The common measure of dispersion is the standard deviation, which is a measure of the breadth of values possible for the variable. Normally, the larger the standard deviation, the greater the relative risk. Finally, its shape or the type of distribution may distinguish a probability variable. Distribution shapes that are commonly continuous distributions used in project risk analysis are the normal distribution, the lognormal distribution, and the triangular distribution.

All four distributions have a single high point (the mode) and a mean value that may or may not equal the mode. Some of the distributions are symmetrical about the mean, whereas others are not. Selecting an appropriate probability distribution is a matter of which distribution is most like the distribution of actual data. In cases where insufficient data is available to completely define a probability distribution, one must rely on a subjective assessment of the needed input variables.

The type of outputs a technique produces is an important consideration when selecting a risk analysis method. Generally speaking, techniques that require greater rigor, demand stricter assumptions, or need more input data generally produce results that contain more information and are more helpful. Results from risk analyses may be divided into three groups according to their primary output:

1. Single parameter output measures
2. Multiple parameter output measures
3. Complete distribution output measures

The type of output required for an analysis is a function of the objectives of the analysis. If, for example, a project manager needs approximate measures of risk to help in project selection studies, simple mean values (a single parameter) or a mean and a variance (multiple parameters) may be sufficient. On the other hand, if a project manager wishes to use the output of the analysis to aid in assigning contingency to a project, knowledge about the precise shape of the tails of the output distribution or the cumulative distribution is needed (complete distribution measures). Finally, when identification and subsequent management of the key risk drivers are the goals of the analysis, a technique that helps with such sensitivity analyses is an important selection criterion.

Sensitivity analysis is a primary modeling tool that can be used to assist in valuing individual risks, which is extremely valuable in risk management and risk allocation support. A "tornado diagram" is a useful graphical tool for depicting risk sensitivity or influence on the overall variability of the risk model. Tornado diagrams graphically show the correlation between variations in model inputs and the distribution of the outcomes; in other words, they highlight the greatest contributors to the overall risk. Figure 2.3 is a tornado diagram for a sample project. The length of the bars on the tornado diagram corresponds to the influence of the items on the overall risk.

The selection of a risk analysis method requires an analysis of what input risk measures are available and what types of risk output measures are desired. These methods range from simple, empirical methods to computationally complex, statistically based methods.

Traditional methods for risk analysis are empirically developed procedures that concentrate primarily on developing cost contingencies for projects. The methods assign a risk factor to various project elements based on historical knowledge of relative risk of various project elements. For example, documentation costs may exhibit a low degree of cost risk, whereas labor costs may display a high degree of cost

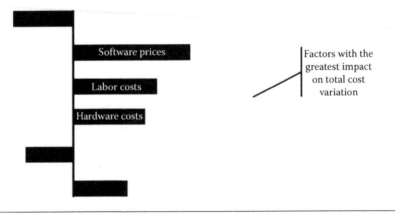

**Figure 2.3**  A tornado diagram.

risk. Project contingency is determined by multiplying the estimated cost of each element by its respective risk factors. This method profits from its simplicity and does produce an estimate of cost contingency. However, the project team's knowledge of risk is only implicitly incorporated in the various risk factors. Because of the historical or empirical nature of the risk assessments, traditional methods do not promote communication of the risk consequences of the specific project risks. Likewise, these techniques do not support the identification of specific project risk drivers. These methods are not well adapted to evaluating project schedule risk.

Analytical methods, sometimes called second-moment methods, rely on the calculus of probability to determine the mean and standard deviation of the output (i.e., project cost). These methods use formulas that relate the mean value of individual input variables to the mean value of the variables' output. Likewise, there are formulas that relate the variance (standard deviation squared) to the variance of the variables' output. These methods are most appropriate when the output is a simple sum or product of the various input values. The following formulas show how to calculate the mean and variance of a simple sum.

For sums of risky variables, $Y = x1 + x2$. The mean value is $E(Y) = [E(x1) + E(x2)]$. The variance is sigma sub $Y$ squared = sigma sub x1 squared + sigma sub x2 squared.

For products of risky variables, $Y = x1 * x2$. The mean value is $E(Y) = [E(x1) * E(x2)]$. The variance is sigma sub $Y$ squared = ($E(x1)$ squared * sigma sub x2 squared) + ($E(x2)$ squared * sigma sub x1 squared) + (sigma sub x1 squared * sigma sub x2 squared).

Analytical methods are relatively simple to understand. They require only an estimate of the individual variable's mean and standard deviation. They do not require precise knowledge of the shape of a variable's distribution. They allow specific knowledge of risk to be incorporated into the standard deviation values. They provide for a practical estimate of cost contingency. Analytical methods are not particularly useful for communicating risks; they are difficult to apply and are rarely appropriate for scheduled risk analysis.

Simulation models, also called Monte Carlo methods, are computerized probabilistic calculations that use random number generators to draw samples from probability distributions. The objective of the simulation is to find the effect of multiple uncertainties on a value quantity of interest (such as the total project cost or project duration). Monte Carlo methods have many advantages. They can determine risk effects for cost and schedule models that are too complex for common analytical methods. They can explicitly incorporate the risk knowledge of the project team for both cost and schedule risk events. They have the ability to reveal, through sensitivity analysis, the impact of specific risk events on the project cost and schedule.

However, Monte Carlo methods require knowledge and training for their successful implementation. Input to Monte Carlo methods also requires the user to know and specify exact probability distribution information, mean, standard deviation, and distribution shape. Nonetheless, Monte Carlo methods are the most common for project risk analysis because they provide detailed, illustrative information about risk impacts on the project cost and schedule.

Monte Carlo analysis histogram information is useful for understanding the mean and standard deviation of analysis results. The cumulative chart is useful for determining project budgets and contingency values at specific levels of certainty or confidence. In addition to graphically conveying information, Monte Carlo methods produce numerical values for common statistical parameters, such as the mean, standard deviation, distribution range, and skewness.

Probability trees are simple diagrams showing the effect of a sequence of multiple events. Probability trees can also be used to evaluate specific courses of action (i.e., decisions), in which case they are known as decision trees. Probability trees are especially useful for modeling the interrelationships between related variables by explicitly modeling conditional probability conditions among project variables. Historically, probability trees have been used in reliability studies and technical performance risk assessments. However, they can be adapted to cost and schedule risk analysis quite easily. Probability trees have rigorous requirements for input data. They are powerful methods that allow the examination of both data and model risks. Their implementation requires a significant amount of expertise; therefore, they are used only on the most difficult and complex projects.

**Risk Checklists**

Checklist 1, Framework for Project Plan, sets forth the key aspects of project implementation that need to be addressed and the important issues that need to be considered for each aspect. To help mangers consider the wide variety of risks any project could face, Checklist 2, Examples of Common Project-Level Risks, sets forth examples of major areas in which risks can occur and examples of key risks that could arise in each area.

Monitoring will be most effective when managers consult with a wide range of team members and, to the maximum extent possible, use systematic, quantitative data on both implementation progress and project objectives. Checklist 3, Ongoing Risk Management Monitoring for Projects, provides a useful framework for ongoing risk management monitoring of individual projects. Checklist 4, Are Risks Adequately Addressed in Project Plan?, is useful for ensuring that risks are discussed in detail.

**Checklist 1**   Framework for Project Plan

| | |
|---|---|
| Project | |
| Responsible manager | |
| Mission | Articulate clearly the mission or goal/vision for the project. |
| Objectives | Ensure that the project is feasible and will achieve the project mission. Clearly define what you hope to achieve by executing the project and make sure project objectives are clear and measurable. |
| Scope | Ensure that an adequate scope statement is prepared that documents all the work of the project. |
| Deliverables | Ensure that all deliverables are clearly defined and measurable. |
| Milestones/costs | Ensure that realistic milestones are established and costs are properly supported. |
| Compliance | Ensure that the project meets legislative requirements and that all relevant laws and regulations have been reviewed and considered. |
| Stakeholders | Identify team members, project sponsor, and other stakeholders. Encourage senior management support and buy-in from all stakeholders. |
| Roles and responsibilities | Clarify and document roles and responsibilities of the project manager and other team members. |
| Work breakdown structure (WBS) | Make sure that a WBS has been developed and that key project steps and responsibilities are specified for management and staff. |

| | |
|---|---|
| Assumptions | Articulate clearly any important assumptions about the project. |
| Communications | Establish main channels of communications and plan for ways of dealing with problems. |
| Risks | Identify high-level risks and project constraints and prepare a risk management strategy to deal with them. |
| Documentation | Ensure that project documentation will be kept and is up to date. |
| Boundaries | Document specific items that are not within the scope of the project and any outside constraints to achieving goals and objectives. |
| Decision-making process | Ensure that the decision-making process or processes for the project are documented. |
| Signatures | Key staff signature sign off. |

**Checklist 2**   Examples of Common Project-Level Risks

| CATEGORY | RISK |
|---|---|
| Scope | Unrealistic or incomplete scope definition |
| | Scope statement not agreed to by all stakeholders |
| Schedule | Unrealistic or incomplete schedule development |
| | Unrealistic or incomplete activity estimates |
| Project management | Inadequate skills and ability of the project manager |
| | Inadequate skills and ability of business users |
| | Inadequate skills and ability of vendors |
| | Poor project management processes |
| | Lack of or poorly designed change management processes |
| | Lack of or poorly designed risk management processes |
| | Inadequate tracking of goals/objectives throughout the implementation process |
| Legal | Lack of legal authority to implement project |
| | Failure to comply with all applicable laws and regulations |
| Personnel | Loss of key employees |
| | Low availability of qualified personnel |
| | Inadequate skills and training |
| Financial | Inadequate project budgets |
| | Cost overruns |
| | Funding cuts |
| | Unrealistic or inaccurate cost estimates |
| Organizational/ business | Lack of stakeholder consensus |
| | Changes in key stakeholders |
| | Lack of involvement by project sponsor |
| | Loss of project sponsor during project |

| | |
|---|---|
| | Changes in office leadership |
| | Organizational structure |
| Business | Poor timing of product releases |
| | Unavailability of resources and materials |
| | Poor public image |
| External | Congressional input or interest |
| | Changes in related systems, programs, etc. |
| | Labor strikes or work stoppages |
| | Seasonal or cyclical events |
| | Lack of vendor and supply availability |
| | Financial instability of vendors and suppliers |
| | Contractor or grantee mismanagement |
| Internal | Unavailability of business or technical experts |
| Technical | Complex technology |
| | New or unproven technology |
| | Unavailability of technology |
| Performance | Unrealistic performance goals |
| | Immeasurable performance standards |
| Cultural | Resistance to change |
| | Cultural barriers or diversity issues |
| Quality | Unrealistic quality objectives |
| | Quality standards unmet |

**Checklist 3**   Ongoing Risk Management Monitoring for Projects

REVIEW PERIOD: _____*

Section 1: Progress and Performance Indicators

| PROJECT IMPLEMENTATION OR OUTCOME OBJECTIVE | PROGRESS/ PERFORMANCE INDICATOR | STATUS OF INDICATOR | ARE ADDITIONAL ACTIONS NEEDED? | NOTES |
|---|---|---|---|---|
| A | | | | |
| B | | | | |
| C | | | | |
| D | | | | |

Section 2: Reassessment of Risks

| IDENTIFIED RISK | ACTIONS TO BE TAKEN | STATUS AND EFFECTIVENESS OF ACTIONS | ARE ADDITIONAL ACTIONS NEEDED? | NOTES |
|---|---|---|---|---|
| 1 | | | | |
| 2 | | | | |
| 3 | | | | |
| 4 | | | | |

*Managers should establish time frames for periodic reviews in addition to ongoing monitoring of program data.

**Checklist 4**   Are Risks Adequately Addressed in Project Plan?

| | PROJECT DESIGN | | PROJECT IMPLEMENTATION | | |
|---|---|---|---|---|---|
| RISK MANAGEMENT ACTION | YES | NO | YES | NO | COMMENTS |
| In developing the project plan, were stakeholders and experts outside the responsible project office consulted about their needs? | | | | | |
| Does the project plan address both internal and external hazards that could impede implementation or performance (see Checklist 2)? | | | | | |
| • Have all relevant laws and regulations been considered? | | | | | |
| • Have all safety/security concerns been considered (patient safety, animal safety, data and property security, etc.)? | | | | | |
| Has a strategy been implemented to prevent or mitigate all identified risks? | | | | | |
| Is reliable, up-to-date data available to allow tracking of project implementation and performance so that problems can be identified early? | | | | | |
| • If not, has an expectation been set that this will be done? | | | | | |
| Are expectations clear and reasonable for the project and for each team member (what, when, and how) and consistent with available resources? | | | | | |
| Are mechanisms in place to ensure effective communication with responsible officials—both within the team and with other stakeholders as necessary? | | | | | |
| If problems occur, can decisions be made quickly? | | | | | |
| Does project have clear goals and objectives that are being continually tracked to ensure they are being achieved? | | | | | |
| Is there a clear statement of how the new process/system will be an improvement over the current process/ system? | | | | | |

Is there clear and accurate baseline
  data for comparing the new process to
  the old process?
Is there a lessons learned component so
  we will be able to use and share the
  good and bad lessons from the project?

## Conclusion

Risk is inherent in all projects. The key to project success is to identify risk and then deal with it. Doing this requires the project manager to identify as many risks as possible, categorize those risks, and
then develop a contingency plan to deal with each risk. Project plans
should always contain a risk analysis.

# 3

# RISK ANALYSIS TECHNIQUES

In Chapter 2 we provided an overview of dealing with risk, which is inherent in all information technology (IT) projects. Doing this requires the project manager to identify as many risks as possible, categorize those risks, and then develop a contingency plan to deal with each risk. Project plans should always contain a risk analysis.

**What Is Risk?**

*Project risk* is defined as an uncertain event or condition that, if it occurs, has a positive or negative effect on a project objective. As implied in the definition, the source of risk is uncertainty. Although uncertainty may, in some cases, be internal to the project (e.g., lack of accountability, poor leadership, or poor teamwork), the greatest source of uncertainty is in the external context (everything outside of the project).

*Risk management* is defined as the systematic process of identifying, analyzing, and responding to project risk. Generally, risk management consists of risk assessment (evaluating the risks) and risk control (acting to avoid, lessen, or prevent the risk). The subprocesses of risk assessment include risk acknowledgement, risk identification, risk analysis, and risk prioritization. The subprocesses of risk control include risk management planning, risk resolution, and risk monitoring.

It is useful to distinguish between two categories of risks: known risks and unknown risks. Those risks that have been identified and analyzed can be classified as *known risks*. It is possible to manage known risks. Within this category of known risks are two risk subtypes:

- Risks that are opportunities—Some risks represent opportunities for the project. Such risks are favorable and may be pursued to the benefit of the project's objectives.
- Risks that are threats—Some risks represent threats to the project. Such risks are unfavorable but may be accepted if they are in balance with the reward that may be gained by taking the risk.

All other risks (risks that have not been identified and analyzed) can be classified as **unknown risks**. Unknown risks cannot be managed. Such risks are often addressed by applying a general contingency margin (e.g., schedule margin, budget margin) based on past experience with similar projects.

### Risk Management Planning Tools

The process of deciding how to approach and plan the risk management activities for a project is called risk management planning. The appropriate amount of risk management planning will be determined by considering the nature of the risk and the importance of the project. For example, a project involving life-threatening risk (e.g., manned space launch) requires a higher level of risk management planning. Similarly, an information technology project of critical importance to the organization's ability to continue to conduct operations (i.e., a mission critical project) will require a higher level of risk management planning.

Risk management planning attempts to determine the level, type, and visibility of a risk management activity based on the nature of the risk and the importance of the project. Risk management activities must be planned to occur throughout the project.

The *risk management plan* is a subcomponent of the overall project plan. The risk management plan describes the overall approach to managing risk. It describes the process by which risks will be identified, analyzed, monitored, and controlled.

Risk identification involves determining which risks might affect the project and documenting their characteristics. Risk identification is an iterative process usually performed by the project team. In some cases, the process may be performed by a dedicated risk management team.

The primary output of the risk identification process is a listing of possible project risks. For ease of risk planning and control, identified risks should be sorted into risk categories. Such categories should reflect common sources of risk for the industry or application area. Risk categories and some associated project risks might include:

- Technical risk—New technology, unproven technology, complex technology, and changes in the technology during the project
- Performance risk—Unrealistic performance goals, unrealistic budget, and unrealistic schedule
- Project management risk—Inadequate quality in the project plan, poor project management discipline, poor allocation of time and resources, inadequate/interrupted funding, and poorly defined customer requirements
- Organizational risks—Resource conflicts with other projects, lack of top management support, and top management's failure to acknowledge the higher risks inherent in project work
- External risks—Labor unrest, country risk, weather risk, and changing customer priorities

Another output of the risk identification process is a set of risk triggers. These are sometimes called risk symptoms or warning signs. They are an indication that a risk has occurred or is about to occur. Examples of risk triggers would be failure to meet an important project milestone, a significant budget overrun, or loss of a key resource.

Tools that may be employed in the risk identification process include documentation reviews, brainstorming, Delphi technique, interviewing, assumption analysis, and system-level diagrams.

*Documentation Reviews*

The initial step in risk identification is usually a documentation review by the project team. The existing project plan and prior project files are reviewed to identify sources of risk. The existing project plan and all documentation generated during preproject activities (e.g., needs assessment, feasibility study, project proposal, project contract, statement of work) need to be carefully examined to identify anticipated project risks.

### Brainstorming

Brainstorming is a frequently used risk identification technique. It is a creative process in which the project team or a multidisciplinary set of experts engages in idea generation, focused on identification of possible project risks.

### Delphi Technique

The Delphi technique is a group decision-making technique that is intended to insulate members from the undue influence of others, thereby producing better decisions. The Delphi technique uses a series of anonymous questionnaires to reach a consensus.

### Interviewing

Risks can be identified through interviews with project stakeholders, experienced project managers, and subject matter experts. The interviewees should attempt to identify possible project risks based upon this information and their experience with similar projects.

### Assumptions Analysis

An important element of risk identification is to identify the underlying assumptions, determine the validity of those assumptions, and identify any project risks related to these assumptions.

### System-Level Diagrams

Projects are frequently engaged in the development of complex systems (e.g., new products, information systems, manufacturing systems). Risks are often manifested in the relationships between system components. Certain system-level diagrams (e.g., cause-and-effect diagrams, system relationship diagrams, interface diagrams, and influence diagrams) document these relationships and can be quite useful in identifying related risks.

### Risk Impact

Qualitative risk analysis is defined as the process of assessing the impact and likelihood of identified risks. As implied by this definition, each specific risk event has two dimensions: the risk impact (sometimes called consequences) and the risk's likelihood of occurrence (risk probability).

Risk impact is the effect on project objectives if the risk event occurs. Risk impact is typically measured on a risk impact scale. Such scales may be ordinal scales (in order according to rank such as low, medium, high) or they may be cardinal scales (having values such as 1, 2, 3).

Values on a cardinal scale may be linear (e.g., 2, 4, 6, 8, 10) or they may be nonlinear (e.g., 2, 4, 8, 16, 32). Using a nonlinear risk impact scale would reflect a company's desire to avoid high-impact risks. For ease of calculation, risk impact may also be expressed on a scale from 0.0 (no impact) to 1.0 (most severe impact).

Decisions about how risk impact will be measured need to be made as early as possible during the project life cycle. A risk impact rating matrix is a very effective way to describe different levels of risk as they apply to specific project objectives. In Figure 3.1, a three-point impact scale (low, medium, high) was used in distinguishing between various levels of risk impact.

| Risk impact rating matrix | | | |
|---|---|---|---|
| Project objective | Low impact | Medium impact | High impact |
| Cost | <5% cost increase | 5%–20% cost increase | >20% cost increase |
| Schedule | <5% schedule slip | 5%–20% schedule slip | >20% schedule slip |
| Scope | <25 change request | 25–50 change requests | >50 change requests |
| Quality | Internally detected errors<br><br>Known cause | Internally detected errors<br><br>Unknown cause | Customer detected errors |

**Figure 3.1**   Risk impact rating matrix.

*Risk probability* (the probability of the risk occurring) is usually measured on a scale that ranges between 0.0 (impossibility) and 1.0 (certainty). Risk probabilities are also sometimes expressed as a percentage, with 0% being impossibility and 100% being certainty.

Risk probabilities are typically difficult to establish with any precision. Experience with similar projects, experience with similar risk events, expert judgment, lessons-learned documents, and historical data can all be useful in establishing more accurate risk probabilities.

By multiplying the numerical risk impact rating (using a cardinal scale of 0.0 to 1.0 for impact severity) and the risk probability (a cardinal number from 0.0 to 1.0 expressing the likelihood that the risk will occur), the risk score for a specific risk can be calculated.

$$\text{Risk impact} \times \text{Risk probability} = \text{Risk score}$$

The risk scoring matrix in Figure 3.2 shows example values for probability, risk impact, and the calculated risk scores. It also illustrates the assignment of a risk category (low, medium, high) based on the numeric value of the calculated risk scores. Cell colors (white, light gray, dark gray) indicate the respective risk category (low, medium, high) that has been assigned to risks with a specific risk score. Risk scores should be determined for all of the risks identified during the risk identification process.

| Risk scoring matrix for a specific risk (Probability x impact) | | | |
|---|---|---|---|
| **Impact on organization** | | | |
| **Probability** | **Minimal impact** 0.2 | **Mild impact** 0.4 | **Moderate impact** 0.6 | **Strong impact** 0.8 |
| 0.9 | .18 | .36 | .54 | .72 |
| 0.7 | .14 | .28 | .42 | .56 |
| 0.5 | .10 | .20 | .35 | .40 |
| 0.3 | .06 | .12 | .18 | .24 |
| 0.1 | .02 | .04 | .06 | .08 |

| Low risk | | Medium risk | | High risk |
|---|---|---|---|---|

**Figure 3.2**   Risk scoring matrix.

A previously identified risk that occurs during the performance of the project is known as a contingency. Formal risk management activities involve the development of contingency plans (i.e., risk response), which are plans to be activated if specified risks occur.

Assumptions that have been made, such as availability of funding by a certain date or availability of a key resource, present a significant source of risk in projects. Therefore, an important aspect of qualitative risk analysis is the testing of project assumptions. Assumptions must be tested in two dimensions: testing the stability of the assumptions (the probability that the assumption will continue to be true over an extended period of time) and determining the consequences if the assumption later proves to be false. For each assumption being tested, alternative assumptions that may be true should be identified. The alternative assumptions should be evaluated in terms of their consequences for the project if they are later determined to be true.

As an example of the impact assumptions can have, consider the International Space Station (ISS) program. The program was launched in 1984 with the assumption that Congressional support and funding would be continuously available as planned over the 15-plus years of the program. However, the assumption regarding long-term availability of Congressional support and funding later proved to be false.

### More on Quantitative Risk Analysis

You were introduced to quantitative risk analysis in Chapter 2. Let's delve a bit more into this topic. Quantitative risk analysis seeks to quantify the risk exposure for a single risk or for an entire project. It assists in identifying the risks that require the most management attention. Quantitative risk analysis usually follows qualitative risk analysis.

The quantitative risk analysis process requires various inputs, such as risk management plan, list of prioritized risks, and expert judgment. These inputs are transformed through various techniques, such as sensitivity analysis and decision tree analysis.

There are many types of risks, such as technical, cost, schedule, performance, marketing, and financial. The financial consequence of a given risk occurring is known as risk exposure. There is a risk exposure associated with each type of risk. The total project risk exposure can be calculated by summing the risk exposures for all risk types.

To calculate the risk exposure for a specific risk type, it is necessary to first estimate two factors:

- Risk probability—What is the probability, expressed on a scale from 0.0 to 1.0, that the risk will occur?
- Risk impact—What are the financial consequences, expressed in dollars, if the risk occurs?

Using these two factors, the risk exposure (in dollars) for any risk type can be calculated as follows:

$$(\text{Risk probability})_{\text{Type}} \times (\text{Risk impact})_{\text{Type}} = (\text{Risk exposure})_{\text{Type}}$$

To use marketing as a general example, total marketing risk would be calculated as follows:

$$(\text{Risk probability})_{\text{Marketing}} \times (\text{Risk impact})_{\text{Marketing}} = (\text{Risk exposure})_{\text{Marketing}}$$

As a specific example of calculating the risk exposure for marketing, consider the following. If the $(\text{Risk probability})_{\text{Marketing}}$ was estimated at 0.1 and the $(\text{Risk impact})_{\text{Marketing}}$ was estimated to be \$500,000, then, the marketing risk exposure (\$50,000) would be calculated as follows:

$$0.1 \times \$500,000 = \$50,000$$

The risk exposure for each identified risk must be calculated in this manner. The total project risk exposure (expressed in dollars) can then be calculated by summing the risk exposures across all risk types:

$$RE_{\text{Technical}} + RE_{\text{Cost}} + RE_{\text{Schedule}} + RE_{\text{Performance}} + RE_{\text{Marketing}} + RE_{\text{Financial}}$$
$$= RE_{\text{Project}}$$

*Sensitivity Analysis*

Sensitivity analysis is the study of how changes in the coefficients (the various risk probability factors) of an equation will affect the calculated risk exposure for the project.

Sensitivity analysis can proceed only after the project risk equation has been solved (all the factors have been estimated and the $RE_{Project}$ has been calculated). The $RE_{Project}$ calculated at this time will be our baseline project risk exposure.

Our approach to sensitivity analysis will involve holding all of the risk probability factors constant, except for one factor that will be systematically varied. Each time that we vary the chosen risk probability factor, we will recalculate $RE_{Project}$ and note the change from its baseline value. In some cases, sensitivity analysis will reveal coefficients that have a dramatic effect on the $RE_{Project}$. In sensitivity analysis, we methodically cycle through all of the coefficients of the equation—changing the selected coefficient, holding the other coefficients constant, and recording the impact on $RE_{Project}$. If this process uncovers risk probability factors that have considerable leverage on the total project risk, this risk category can be targeted for extra management attention.

*Decision Analysis*

When a decision maker is faced with several alternatives and an uncertain or risk-filled future, an approach called decision analysis can be used to determine optimal strategies. The decision analysis approach involves:

- Identifying the alternatives (the decisions)
- Identifying the future events that might occur (the states of nature)
- Calculating the expected values for every possible combination of decisions and states of nature

When the number of outcomes is reasonably small, a technique known as a decision tree can be used to guide decision making. As an example, consider the decision of whether to go to the movies or stay home and watch TV, as shown in Figure 3.3. The situation can be illustrated with a decision tree in which a rectangle represents a decision and a circle with related branches represents the events and the possible outcomes (or states of nature).

The simple example represents a state of near certainty. After all, we could look at movie reviews on Fandango or some social media websites to provide information that would allow us to predict the outcomes with considerable certainty.

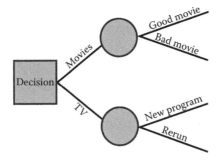

**Figure 3.3**   Decision tree.

### Risk Response

Risk response planning seeks to develop an agreed upon risk response for each identified risk and to identify individuals to take ongoing responsibility for managing the risk response. Several possible risk response strategies are available: avoidance, transference, mitigation, and acceptance. Both primary and backup strategies may be needed for each identified risk.

Risk avoidance involves eliminating a risk or protecting the project objectives from the impact of the risk. Generally, the more that we know about the project and its related risks, the better we are able to protect against those risks. Previous projects, expert opinion, analysis, and a collaborative decision-making process are sources of information that can inform our risk avoidance efforts.

Risk transference seeks to transfer risks to another party and to give the other party the responsibility for the risk response. Risk transference does not eliminate risk; it simply gives another party responsibility for managing the risk.

Risk transference is most effective in transferring financial risk. Insurance, performance bonds, warranties, and guarantees are conventional methods of risk transference. Such risk transfers usually involve the payment of a risk premium to the party assuming the risk.

In the context of project management, risk mitigation may involve various actions to reduce the impact if a specific risk should occur. For example, slack resources (having a spare computer), planned redundancy (backing up critical computer data), and cross-training of employees can respectively reduce the impact of the breakdown of a critical device, the loss of critical data, or the untimely loss of a key

employee. Although there is additional cost and complexity involved with slack resources and redundancy, the risk severity sometimes justifies such risk mitigation efforts.

Risk acceptance involves the decision to leave the project plan unchanged and not to implement any risk avoidance, transference, or mitigation strategies with respect to a specific risk. Passive acceptance of the risk leaves the project team to deal with the risk, if and when it occurs. Each risk that is actively accepted should have a contingency plan specifying "what we will do if this risk occurs." The most common risk acceptance response is to establish a management reserve (or contingency allowance) to compensate for known risks that have been accepted. The management reserve may be in the form of time, money, or resources.

The primary output of the risk response planning process is a formal risk response plan. This plan should be written in sufficient detail that it clearly specifies what actions are to be taken, when the actions should be taken, and by whom the actions should be taken.

The risk response plan should include:

Identified risks
  • Descriptions of all identified risks
  • Areas of the project affected
  • Causes of the risks
  • How project objectives might be impacted
Risk responsibilities
  • Assigned risk owners for each risk (the name of responsible individuals)
  • Assigned risk responsibilities for each risk
Risk analyses
  • Results of the qualitative analyses for all identified risks
  • Results of the quantitative analyses for all identified risks
Risk responses
  • Agreed upon risk strategies (avoidance, transference, mitigation, or acceptance) for each risk
  • Supporting rationale (Why was this risk strategy chosen?)
Contingency plans
  • Contingency plans for each accepted risk
    – Defined risk triggers

- Assigned responsibility (the names of responsible individuals) to monitor for occurrence of the risk trigger
- Residual risk expected to remain after the contingency plan is executed
- Specific and detailed actions needed to implement the contingency plan
- Allocated budget for funding the contingency plan effort
- Schedule for completing the contingency plan
- Fallback plan (for high-impact risks)

**Risk Monitoring and Control**

Risk monitoring is a passive process of tracking risks, the risk response plan, and specific risk response actions being executed. Risk control is an active process for managing and controlling project risk. Risk control activities may include choosing alternative strategies, implementing a contingency plan, taking corrective action, or replanning the project. Both risk monitoring and risk control are ongoing processes.

The tools and techniques used in risk monitoring and control are:

- Risk response audits—Periodic examinations by appointed risk auditors. These audits seek to examine and document the effectiveness of risk responses and risk owners.
- Risk reviews—Regularly scheduled reviews of risk ratings and risk priorities that may change during the life of the project.
- Technical performance measurement—A comparison of actual technical accomplishments with the planned technical accomplishments. Failure to demonstrate a planned level of technical performance at a project milestone signals a significant risk event that may impact project goals.
- Additional risk response planning—Planning required as a result of newly emergent risks or risks with greater impact on objectives than expected.

The primary outputs of the risk monitoring and control process are:

- Workaround plans—Unplanned responses to emerging risks that were previously unidentified or accepted.

- Corrective actions—The actual performance of contingency plans or workaround plans.
- Project change requests—Formal requests to change the project plan in response to specific risks.
- Updates to the risk response plan—Reassessments of risks, risk rankings, and risk response actions.
- Risk database—A repository of information that has been gathered and used in the risk management process.
- Updates to the risk identification checklist—Experience-based modifications and improvements to the standardized risk identification checklist.

### Conclusion

Understanding project risk is a key component of project planning. It is important to understand that there are many categories of risks (e.g., business, environment, product, employee). A variety of tools can be used to assist in the identification of risks. Once all risks are identified, assessed in terms of probability and impact, a contingency plan can be developed to mitigate the specific risk. A risk plan should not be developed and filed away. Risks must be actively monitored if they are to be successfully controlled.

# 4

## Auditing Knowledge Management to Reduce Risk

It's no secret that computers make mistakes; or rather the programs have been written with misconceptions and errors in the code. This has the very real side effect of dramatically increasing risk.

In very large systems, especially knowledge-based systems, where the code is complex and the programs are very long, the possibility for error is very high. Electronic data processing (EDP) auditors are charged with ensuring that the process of creating and running of these advanced computer systems is consistent with recognized development practices.

### Complexity

Knowledge-based systems are both harder and easier to audit than your garden-variety computer system. The wind can blow in either direction depending upon what building tools are used. When you build a log cabin, you clear the land and lay the foundation. Log after log is joined together inching upward toward the sky. When it is done you can walk about, inside and out, rubbing your finger along the neatly stacked logs. You can see at a glance just how the log cabin was made.

Then there is a skyscraper. Acres of dirt are hauled away to lay the foundation, often as deep as a dungeon. Steel girders are hoisted heavenward by groaning cranes. Hundreds of hard-hatted workers scurry around on the high crossbeams. The skyscraper rises as does the log cabin—the difference is the complexity. Many knowledge-based systems are built to be as easily understood as our log cabin. And many are built to skyscraper levels of complexity. And many more are built somewhere between the two extremes. Three factors determine a system's complexity.

The first factor is tool complexity. There are hundreds of tools in use today, from programming languages to expert system shells to data mining tools to machine learning and big data analytical tools. Many of these tools are simple and self-documenting. But many more require great deal of training and expertise to build effective systems.

The second factor is the level of expertise being snared in the fishing net. A system that makes wine selections is simpler than a system that determines when your false teeth will be ready, is simpler than a system that makes recommendations about personal investment, is simpler than a system that authorizes loans, is simpler than a system that makes buy-sell decisions for foreign exchange deals, is simpler than a system that authorizes the launch of the space shuttle.

There are four different categories of knowledge-based systems: A *knowledge-intensive* system is knowledge-bound but uses a simple computing environment and usually acts in an advisory capacity. The other extreme is the *technology-intensive* system, which contains limited knowledge or knowledge in a limited domain, but requires advanced computing prowess. This sort of system can usually be found in areas where improvement in organizational productivity is envisioned. The most exotic type of system is a strategic impact system, where not only complex knowledge is encoded but the system is technologically complex. On the low-end of the totem pole are *personal productivity* systems where limited amounts of knowledge as well as simple technology are the characteristics. The high end is exemplified by foreign exchange systems where advanced workstations are used, massive data feeds are integrated, and complex rules of knowledge are encoded. Another example would be the massive Wal-Mart data mining and warehousing system.

The third factor that determines a system's complexity is the most vital and the most problematic. Will we adhere to the decisions of the knowledge-based system?

### Audit Questions

Since knowledge-based systems are a level of complexity greater than a conventional system, the audit should be a level of complexity greater than the conventional system audit.

*Where Did You Get the Knowledge for This System?*

Expert knowledge should come from an expert. No compromise here; no "Let's use Mary because she's here." The audit process should check to ensure that the best possible candidate was chosen.

It is substance that is at the heart of a knowledge-based system. So how do we get it? Unfortunately, for many knowledge-based systems, the expert selected is the person who has the most free time. For other systems, the top guru in the department is selected, who is then taken offline for several weeks or months in an attempt to wrest the knowledge from its very roots. Some very smart systems use the combined expertise of several experts, which gives the ability to incorporate differing or competing viewpoints. Perhaps the most brilliant of systems incorporate the views and opinions of experts external to the company itself. The degree of expertise selected should be proportional to the level of strategic competitiveness the system is expected to display.

Functions within an organization are often graded proportionally to their worth to the organization. In a bank, those responsible for approving mortgage loans are a higher grade within the organization then those who authorize credit card purchases. The organization inherently recognizes the complexity of the mortgage authorization process as compared to the credit card authorization process. American Express, in its development of its own credit authorization knowledge-based system for its green card, appropriately used the senior authorizer as the basis for the rules governing the very successful system. The rules governing this sort of credit approval may be many but are noncomplex and based on a moderate number of variables. Mortgage approval is a horse of a different color. Many variables come into play here, some vague or fuzzy. This process must take into account such things as the amount of the mortgage, down payment, security, primary income, secondary income, rate basis, rate spread, terms of the loan, and the list goes on. There is obviously much more risk associated with this process. The seedlings of this brand of knowledge-based system must come from a senior staffer and from more than one staffer to boot.

Most corporations have been quite perceptive about the failings of their own internal organizations and certain areas of expertise. Many made competing organizations for the most experienced personnel.

Many hire big names, such as past members of presidential cabinets, for their unique perspectives of world events. So too, knowledge-based systems. Figure 4.1 makes it readily understandable that there is great risk in building a knowledge-based system that utilizes the wrong level of expertise.

At one prestigious Wall Street firm it was decided to build a trading system that would provide expertise in the time-honored tactic of hedging. Since hedging is more of an art than science, the development team needed to gain the cooperation of the premier participants in this art form. Unfortunately, these extraordinary folks were determined not to part with their knowledge. They felt it was what made them uniquely unmarketable, not an unusual position for an expert to take. Since the development team could not get the cooperation it needed, it turned to textbooks as substitute mentors. Textbooks contain the rules and procedures and not the gut instincts and rules of thumb that would have made this sort of knowledge-based system expert or even brilliant.

As the expertise required continuum in Figure 4.1 shows, there are many gradations of knowledge-based system types, the smarter than conventional systems to the very brilliant. In each of the systems a certain level of expertise is required. If the systems are to be built, management should not look askance at the technology but take an active interest in its success. The way this can be done is to become involved at the point where the actual level of expertise is to be selected. The selection, which can appear trivial, is fraught with risk for those that take a laissez-faire attitude. The continuum in Figure 4.1 clearly shows that some systems are so complex and require as input the analysis of

| Brilliant | Interest rate forecasting | Many high-level outside experts |
|---|---|---|
| Expert | Mortgage processing | Multiple experts, some outside |
| Extremely knowledgeable | Pension check processing | Multiple inside experts |
| Smart | Credit card authorization | Single expert |
| Conventional | Payroll | Little expertise |

**Figure 4.1**  Expertise-required continuum.

so many far-reaching variables, that default to in-house staff as a total solution is not realistic. For these systems, which often reside in key areas of the company, including outside experts to gain a broader perspective is endorsed.

### Did We Get All of the Knowledge?

Once the auditor is confident that a bona fide expert was chosen, a thorough analysis should be done to determine if all the knowledge was captured. How should this be done? One way would be to review the transcripts of the knowledge engineering sessions, but this would be extremely time consuming. Probably the best way is to observe the use of the system in practice and look for the holes. A complete knowledge-based system should leave no question unanswered and no stone unturned. So, it should account for all situations that arise during a sample test period. For example, suppose we build a super-duper credit authorization system. During the system development process the expert worked diligently to encode all the processing rules as well as the expert's gut instinct in approving credit. The system is put into the credit department amid much fanfare. Sixty people come to work Monday morning ready, eager, and willing to work with their new electronic credit adviser. At 9:01 A.M. the phones start ringing off the hook. Real customers. Real credit decisions. Can every situation be handled? And handled correctly?

The auditor should observe this process. A group of credit authorizers should be preselected to log all instances of variances and omissions between what the system does and what they would have done manually. Since this process should also been a part of the original test plan, the auditors should find few problems. Given the enormity of the task of capturing the whole ball of wax, it is a sure thing that the auditors will find some discrepancies.

Auditors enjoy looking for discrepancies and the most common way they find them is with an audit trail. Now audit trails are uncommon to most computer systems for a simple reason. They cost to build and they cost to run. One of the niceties of using some of the knowledge-based technologies is that an audit trail is almost a natural part of system design. So it should be almost trivial for an auditor to perform an after-the-fact audit when this type of technology is used. Here the auditor examines a printed audit trail and compares the system deci-sion to the decision a human credit authorizer would have given.

*How Much Common Sense Does the Knowledge-Based System Exhibit?*

Common sense is hard to systematize and easy to miss. Some experts get so caught up in the esoterics of what they do for a living, they miss the trivial common sensibilities of their daily tasks. In one case a bank built a prototype knowledge-based system for determining the scope of an internal audit. The prototype worked wonderfully in predicting the scope of a local audit for branch bank until it ran aground at one branch that did not have any loans.

*Does the Knowledge-Based System Make Mistakes?*

What happens if the system becomes confused? In the field of psychol-ogy, a confusion matrix is built to determine the ratios between correct responses and those responses that were close but not quite correct. We can use the same technique to determine just how right or wrong a sys-tem is. Let's look at an example. We just built a knowledge-based sys-tem to determine the type of disease based on patients' symptoms. Our list of possible diseases is the flu, a cold, and the black plague. Sitting in one room is a medical expert who examines several patients and makes diagnoses. Across the hall is a computer that also makes diagnoses. Table 4.1 shows a confusion matrix for the differences between the human doctor's diagnosis and computer diagnosis. Remember that in each instance the information given to the doctor was the same as that fed into the system. Presuming that our doctor is a bona fide expert, the responses of the knowledge-based system should exactly parallel those of the doctor. You can see from the confusion matrix that this did hap-pen. Each slot or entry in the matrix represents the percentage of times that the expert system differs from the human expert, which really

**Table 4.1** Confusion Matrix

|              | BLACK PLAGUE | FLU  | COLD |
| ------------ | ------------ | ---- | ---- |
| Black Plague | n/a          | 35%  | 45%  |
| Flu          | 20%          | n/a  | 15%  |
| Cold         | 10%          | 5%   | n/a  |

means the percentage of times the knowledge-based system was confused. For example, when our doctor diagnosed the flu, the knowledge-based system recommended the black plague 20 percent of the time and cold 15 percent of the time. These erroneous conclusions totaled 35 percent of the time. This means that our knowledge-based system was correct 65 percent of the time. The average of these percentages will give us an error threshold that each company must decide is acceptable or not. In the case of our medical experts system, the total error threshold was 57 percent, that is, the system was correct only 57 percent of the time. This would be unacceptable in anybody's medical textbook.

### Is the System Kept Up to Date?

By the sweat of everybody's brow, 3 years and 2000 hours later the system is complete. With great fanfare the system is presented to management followed by a champagne party and many press releases. While everyone is toasting each other and slapping each other on the back, something terrible is happening. The system is becoming obsolete.

You've all heard the old saw about books: As soon as it's in print, it's out of date. Will the same thing hold true about knowledge-based systems? Knowledge-based systems seldom stay static. They grow and expand beyond the developers' wildest dreams.

How do they grow? This is the question that the auditor should ask. Which department has a responsibility for ensuring new tidbits of knowledge? Is it the IT department? The experts? Those that use the knowledge-based system?

And the auditor should be on the look out for adolescent sloppiness. When the system is new there is usually adequate money, time, and enthusiasm to do the job right. After the fanfare is over and the spotlight fades, the system matures into adolescence. Gradually, the heavy hitters move away from the system to something a bit more exciting leaving our little system alone forlorn. When a change has to be made to the system, it will be made during maintenance mode rather than development mode.

Maintenance is not your most interesting work. And those who are assigned to it usually manage to work up the same level of enthusiasm as they do for a weekend trip to the dentist. Changes made in this matter are usually sloppy, or if you are really lucky just careless. The meticulousness exhibited during the development stage, where the team wanted everything to be just so, is usually never replicated. So the auditor would be wise to take a look the procedures put into place to handle the maintenance of these systems.

*Will the Users Actually Use the Knowledge-Based System?*

Most knowledge-based systems are used in an advisory capacity. This means a real live human being will read the decision and do one of two things: use it or ignore it. A well-designed system will track the adherence factor, which can then be used by the auditor to track whether the system is being used effectively. Of course, it is a pretty good idea for management to use this feature, too.

The adherence factor is actually an audit trail of the system's recommendations. For each consultation, the system keeps track of the conclusion and advice. This conclusion can than be compared to the one the staff actually made. To simplify the process, some add an online justification feature that permits the user of the system to enter what was actually done. Those who would serve as auditors will most likely place this high on the list of features to consider.

**An Audit Plan**

Most people think of auditing it as a necessary evil. Systems folks don't think kindly of auditors. No wonder auditors take a fine-tooth comb to every facet of the development process, from file design to documentation to user manuals. Knowledge-based system auditing will pose the following questions:

- Can you re-create an expert system consultation?
- Are the machine learning algorithms flawed in any way?
- Is the data captured during the consultation edited or verified in any way?
- When an error is detected, which users receive immediate notification?
- If an error is detected, does the knowledge base flag the error?
- How rigorously are the answer fields constrained?
- Are new releases of the knowledge base tracked?

- What were the test cases used to verify the system?
- Are new releases quality assured?
- Are new releases endorsed by the expert?
- How are differences between the system and human expert handled?
- Is documentation maintained?

The following are the basic steps in performing the audit:

1. Planning the audit
2. Evaluation of internal controls
3. Audit procedures
4. Completing the audit

The auditor must plan and conduct the audit to ensure the audit risk (reaching an incorrect conclusion based on the audit findings) will be limited to an acceptable level. To eliminate the possibility of assessing audit risk too low, the auditor should perform the following steps:

1. Obtain an understanding of the organization and its environment—The understanding of the organization and its environment is used to assess the risk of material misstatement/ weakness and to set the scope of the audit. The auditor's understanding should include information on the nature of the entity, management, governance, objectives and strategies, and business processes.
2. Identify risks that may result in material misstatements—The auditor must evaluate an organization's business risks (threats to the organization's ability to achieve its objectives). An organization's business risks can arise or change due to new personnel, new or restructured information systems, corporate restructuring, and rapid growth, to name a few.
3. Evaluate the organization's response to those risks—Once the auditor has evaluated the organization's response to the assessed risks, the auditor should then obtain evidence of management's actions toward those risks. The organization's response (or lack thereof) to any business risks will impact the auditor's assessed level of audit risk.
4. Assess the risk of material misstatement—Based on the knowledge obtained in evaluating the organization's responses to

business risks, the auditor then assesses the risk of material misstatements and determines specific audit procedures that are necessary based on that risk assessment.

5. Evaluate results and issue audit report—At this level, the auditor should determine if the assessments of risks were appropriate and whether sufficient evidence was obtained. The auditor will issue either an unqualified or qualified audit report based on their findings.

The auditor evaluates the organization's control structure by understanding the organization's five interrelated control components. They are

1. Control environment—Provides the foundation for the other components. Encompasses such factors as management's philosophy and operating style.
2. Risk assessment—Consists of risk identification and analysis.
3. Control activities—Consists of the policies and procedures that ensure employees carry out management's directions. Types of control activities an organization must implement are preventative controls (controls intended to stop an error from occurring), detective controls (controls intended to detect if an error has occurred), and mitigating controls (control activities that can mitigate the risks associated with a key control not operating effectively).
4. Information and communication—Ensures the organization obtains pertinent information, and then communicates it throughout the organization.
5. Monitoring—Reviewing the output generated by control activities and conducting special evaluations. In addition to understanding the organization's control components, the auditor must also evaluate the organization's general and application controls.

Controls that have an effect on knowledge management processes include:

- Organizational controls—Includes segregation of duty controls.
- Data center and network operations controls—Ensures the proper entry of data into an application system and proper oversight of error correction.

- Hardware and software acquisition and maintenance controls— Includes controls to compare data for accuracy when it is input twice by two separate components.
- Access security controls—Ensures the physical protection of computer equipment, software, and data, and is concerned with the loss of assets and information through theft or unauthorized use.
- Application system acquisition, development, and maintenance controls—Ensures the reliability of information processing.
- Application controls—Application controls apply to the processing of individual accounting applications and help ensure the completeness and accuracy of transaction processing, authorization, and validity. Types of application controls include:
  - Data capture controls—Ensures that all transactions are recorded in the application system, transactions are recorded only once, and rejected transactions are identified, controlled, corrected, and reentered into the system.
  - Data validation controls—Ensures that all transactions are properly valued.
  - Processing controls—Ensures the proper processing of transactions.
  - Output controls—Ensures that computer output is not distributed or displayed to unauthorized users.
  - Error controls—Ensures that errors are corrected and resubmitted to the application system at the correct point in processing.

Application controls may be compromised by the following application risks:

- Weak security
- Unauthorized access to data and unauthorized remote access
- Inaccurate information and erroneous or falsified data input
- Misuse by authorized end-users
- Incomplete processing and/or duplicate transactions
- Flawed algorithms
- Untimely processing
- Communication system failure
- Inadequate training and support

Risk and Knowledge Management

Effective knowledge management is crucial to enterprise risk management (ERM). ERM is based on the fact that business processes, risks, and controls across the organization are interrelated. Effective risk management can only emerge when the organization begins to share and control knowledge systematically across its functions and departments.

Effective ERM guidelines include:

1. Create a risk ownership focus for each employee.
2. Develop a corporate integrity program.
3. Develop a common risk language for the entire organization.
4. Develop a common process classification for the entire organization.
5. Develop a process for managing risks.
6. Internal auditing should become proactively involved in any organizational project or initiative (e.g., key business initiatives, mergers, new business systems, project redesigns).
7. Develop business expertise within the internal audit department. Each auditor should be assigned one aspect of the business and this should become his or her focal point from a knowledge-gathering standpoint.
8. There should be a single focus point for managing the critical business knowledge within an organization. Some organizations have highly structured groups in charge of monitoring, managing, and facilitating the internal sharing of knowledge. Some organizations, however, do not have a group specifically devoted to this task. Many auditors believe that this should be the role of the audit function.

Linking Knowledge Management to Business Performance

It is possible to develop a framework for linking knowledge management (KM) to business performance. To be able to assess the impact of knowledge management, knowledge management initiatives must be aligned to an organization's strategic objectives. One approach is to organize a knowledge management initiative into three stages, which is then further explored through templates supported by detailed guidelines. For each stage, there are steps or thought processes required to structure business problems.

The aim of stage 1 is to provide a structure for formulating a strategic business plan by identifying business drivers, defining strategic objectives or goals, identifying critical success factors, and developing measures for monitoring performance improvement. The outcome of stage 1 is a business improvement plan with performance targets and measurable goals.

An abbreviated template for Stage 1 might consist of:

1.1 Choose a business problem with a knowledge dimension
1.2 Place the business problem in a strategic context by relating it to your external business drivers, strategic objectives, and critical success factors
1.3 Select an appropriate set of measures to monitor progress
1.4 Identify previous, current, target, and benchmark scores for various performance measures

Each of these four steps are supported by detailed guides such as sample performance measures and various glossaries.

The purpose of stage 2 is to determine whether the business problem has knowledge dimensions and to develop specific knowledge management initiatives to address the business problem. A sample template for stage 2 might consist of:

2.1 Clarify the knowledge dimension of your business problem by identifying the KM processes involved
2.2 Develop specific KM initiatives
2.3 Select possible tools to support the KM processes identified
2.4 Identify possible relationships between KM initiatives and performance measures; use a cause-and-effect map for this purpose
2.5 Prepare an action plan and identify change management and resources required

Stage 2 steps are supported by detailed guides such as a questionnaire to identify the knowledge management subprocesses involved, a matrix for the selection of the most appropriate tools, and a checklist to identify possible barriers and facilitators prior to implementation.

In stage 3, a knowledge management evaluation strategy and implementation plan are developed. The output of stages 1 and 2 of the framework is a business improvement strategy underpinned by knowledge management. A sample template for stage 3 might consist of:

3.1 Use the cause-and-effect map developed in 2.4 to assess the likely contribution of the KM initiatives to performance measures

3.2 Assess the probability of success of the KM initiative

3.3 Identify the cost components for implementing each KM initiative and the possible benefits

3.4 Identify the risks for implementing each KM initiative

3.5 Choose an appropriate method to assess the impact of each KM initiative on business performance

3.6 Prioritize KM initiatives based on the measures of performance

Several evaluation techniques may be used:

- Cost minimization analysis—This involves a simple cost comparison of KM initiatives as it is assumed that the outputs are identical or differences between the outputs are insignificant. This technique does not take account of the monetary value of the outputs.
- Risk minimization analysis—The primary goal of a risk minimization analysis is to develop activities that help managers understand important known or potential risks associated with a KM initiative.
- Cost effectiveness analysis—This involves the comparison of KM initiatives where the consequences are measured using the same units.
- Cost utility analysis—This involves a comparison of KM initiatives that are measured in monetary units with the consequences measured using a preference scale (e.g., preference of individuals, teams, or the organization).
- Cost–benefit analysis—This approach provides a comparison of the value if input resources to be used by the KM initiative compared to the value of the output resources the KM initiative might save or create.

Conclusion

In an age of Big Data and smart systems such as IBM's Watson, it is important that the organization carefully assess the risk of its knowledge management endeavors. As we have seen in this chapter, the tools and techniques of auditing are perfect for this purpose.

# 5

# INNOVATION MANAGEMENT TO REDUCE RISK

According to the definition of *innovation* found on *Wikipedia* (http://en.wikipedia.org/wiki/Innovation), innovation is either demand-led or supply-pushed, and this topic is considerably debated. *Wikipedia* is a good example of both definitions of innovation. There was a demand in the marketplace for a free, web-based encyclopedia. The technology of the Internet and the concept of the wiki, a web-application that lets users add and change content (http://en.wikipedia.org/wiki/Wiki), is an excellent example of supply-pushed innovation. The wiki was conceived and developed by Ward Cunningham in the mid-1990s. Steve Lipscomb's World Poker Tour is another example. Poker has taken the world by storm, largely because of Lipscomb's innovative approach to the once seedy concept of the poker tournament.

So, how do you manage to generate this level of innovation and creativity, while simultaneously managing risk?

### Encouraging Innovation

There are two types of innovation. Sustaining advances give our most profitable customers something better, in ways that they define as "better"; and disruptive advances impair or "disrupt" the traditional fashion in which a company has gone to market and made money, because the innovation offers something our best customers *do not* necessarily want. Needless to say that disruptive innovation brings with it a higher degree of risk.

Most software companies continually enhance their line of software products to provide their customers with the features that they have stated they truly desired. This is an example of sustaining innovation. These companies might also strive to come up with products that are radically different from what their customers want in order to expand their base of customers, compete with the competition, or even jump into a completely new line of business. This is an example of disruptive innovation.

Most people equate innovation with a new invention, but it can also refer to a process improvement, continuous improvement, or even new ways to use existing things. Innovation can, and should, occur within every functional area of the enterprise. Good managers are constantly reviewing the internal and external landscape for clues and suggestions about what might come next:

- Research results from R&D (research and development)— One of the challenges is being alert to market opportunities that might be very different than the inventor's original vision.
- Competitors' innovations—Microsoft leveraged Apple's break-through graphical user interface and ultimately became far more dominant and commercially successful than Apple.
- Breakthroughs outside industry.
- Customer requests—A "customer-focused" organization's products and services will reflect a coherent understanding of customer needs. This has been made much easier using social media (e.g., the organization's Facebook and Twitter presence).
- Employee suggestions.
- Trade journals.
- Trade shows and networking.

Some experts argue that a company's product architecture mirrors and is based upon its organizational structure. This is because companies attack that first project or customer opportunity a certain way, and if it works they look to repeat the process and this repetition evolves into a company's "culture." So, when we say a company is "bureaucratic" what we are really saying is that it is incapable of organizing differently to address different customer challenges, because it has been so successful at the original model.

There are a variety of workplace structures that promote innovation. Selecting a representative from the various functional areas (cross-functional teams) and assigning them to solve a particular problem can be an effective way to quickly meld a variety of relevant perspectives and also efficiently pass the implementation stress test, avoiding, for example, the possibility that a particular functional group will later try to block a new initiative and increasing "eyes on" the project, which results in reduced risk. Some variations include a "lightweight project manager" system where each functional area chooses a person to represent it on the project team. The project manager serves primarily as a coordinator. This function is "lightweight" in that the project manager does not have the power to reassign people or reallocate resources. Another alternative is a "tiger team," which consists of individuals from various areas who are assigned and completely dedicated to the project team, often physically moving into shared office space together. This does not necessarily require permanent reassignment, but is obviously better suited for longer-term projects with a high level of urgency within the organization. Finally, some companies have developed innovative partnership models (e.g., cross-company teams or industry coalitions) to share the costs and risks of these high-profile investments.

There are several managerial techniques that can be utilized to spur innovation, as shown in Table 5.1. Before utilizing any one or more of these techniques, the risk of using that technique should be considered. For example, any websites used for research should first be vetted to ensure legitimacy and accuracy of data.

### The Research and Development (R&D) Process

As shown in Figure 5.1, at a very high level every R&D process will consist of

1. Generation of ideas—From the broadest visioning exercises to specific functionality requirements, the first step is to list the potential options (highest degree of risk entertained).
2. Evaluation of ideas—Having documented everything from the most practical to the far fetched, the team can then coolly and rationally analyze and prioritize the components, using agreed-upon metrics and risk factors (risk reduced by careful analysis of risk factors).
3. Product/service design—These "ideas" are then converted into "requirements," often with very specific technical parameters (lowest degree of risk due to careful attention to requirements while considering attendant risks).

**Table 5.1** Promoting Innovation

| TECHNIQUE | DEFINITION/EXAMPLES |
| --- | --- |
| Commitment to problem solving | Ability to ask the "right questions"<br>Build in time for research and analysis |
| Commitment to openness | Analytical and cultural flexibility |
| Acceptance of out-of-the-box thinking | Seek out and encourage different view points, even radical ones |
| Willingness to reinvent products and processes that are already in place | Create a "blank slate" opportunity map, even for processes that appear to be battle-tested and comfortable |
| Willingness to listen to everyone (employees, customers, vendors) | "Open door"<br>Respect for data and perspective without regard to seniority or insider status |
| Keeping informed of industry trends | Constantly scanning business publications/trade journals, and clipping articles of interest<br>"FYI" participation with fellow managers |
| Promotion of diversity, cross-pollination | Forward-thinking team formation, which also attempts to foster diversity<br>Sensitive to needs of gender, race, even work style |
| Change of management policies | Instill energy and fresh start by revising established rules |
| Provision of incentives for all employees, not just researchers/engineers | Compensation schemes to align individual performance with realization of company goals |
| Use of project management | Clear goals and milestones<br>Tracking tools<br>Expanded communication |
| Transfer of knowledge within an organization | Commitment to aggregating and reformatting key data for "intelligence" purposes |
| Provision for off-site teaming | Structured meetings and socialization outside the office to reinforce bonds between key team members |
| Provision for off-site training | Development of individuals through education and experiential learning to master new competencies |
| Use of simple visual models | Simple but compelling frameworks and schematics to clarify core beliefs |
| Use of the Internet for research | Fluency and access to websites (e.g., competitor home pages) |
| Development of processes for implementing new products and ideas | Structured ideation and productization process<br>Clear release criteria<br>Senior management buy-in |
| Champion products | Identify and prioritize those products that represent best possible chance for commercial success<br>Personally engage and encourage contributors to strategic initiatives |

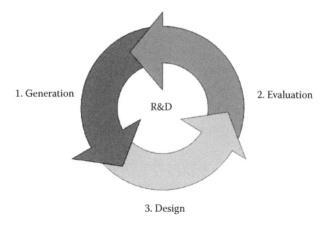

**Figure 5.1** The R&D process.

There are two core elements of this longer-term competency-enhancing work. The first is the generation of ideas. Most companies utilize a standard process to make sure that that everyone has time and motivation to contribute. The second element is to promote an environment conducive to innovation. This includes

1. Cultural values and institutional commitment
2. Allocation of resources
3. Linkage with company's business strategy

Creating an innovation-friendly environment is time consuming and will require the manager to forego focusing on the here and now. When there is constant pressure to hit the numbers or make something happen, it is difficult to be farsighted and build in time for you and your team to create an environment.

Managing innovation is a bit different than creating an environment that promotes innovation. This refers to the service- or product-specific initiative, whether it is a new software product or a streamlined manufacturing process. The big question is, *How do we make this process come together on time and under budget?* There are two main phases to the successful management of innovation. The first phase seeks to stress-test the proposal with a variety of operational, financial, and risk-related benchmarks, such as:

- *Is the innovation "real"?* Is this "next great thing" dramatic enough to justify the costs and risks, financial and otherwise? Does it clearly and demonstrably distance you from your competitors? And can it be easily duplicated once it becomes public knowledge?
- *Can the innovation actually be done? Does the organization have the resources?* This is where you figure out whether the rubber meets the road. You need to ask whether you have the capabilities and functional expertise to realize this vision. Many organizations come up with a multitude of ideas. Upon further examination, they often find that they simply do not have the resources to do the vast majority of them. This might lead them to become innovative in a different way as they search for partners. In other words, some organizations try to couple their brains with someone else's brawn!
- *Is the innovation worth it? Does the innovation fit into the organization's mission and strategic plan?* ROI (return on investment) is the most frequently used quantitative measure to help us plan and assess new initiatives. Probably more useful, however, is ROM (return on management), which poses a fundamental question: What should the CEO and his/her management team focus on? ROM is calculated by first isolating the management value-added of a company, and then dividing it by the company's total management costs.

Return on management =
Management value-added/Management costs

Management value-added is that which remains after every contributor to a firm's inputs gets paid. If management value-added is greater than management costs, you can say that managerial efforts are productive because the managerial outputs exceed managerial inputs. Another way of looking at the return on management ratio is to view it as a measure of productivity. It answers the question of how many surplus dollars you get for every dollar paid for management.

The second phase, design, is the process by which these ideas and concepts get distilled into an actual product design, for example, a website map or a prototype. Many mistakes, which serve to increase risk, are made by delegating this process to lower-level functional experts, when in fact some of these decisions go a long way toward determining the product's ultimate acceptance in the marketplace!

Most of the outward signs of excellence and creativity we associate with the most innovative companies are a result of a culture and related values, which encourage and support managers who use their specific initiatives to also reinforce and strengthen the company's processes. When these processes become "repeatable," they become the rule instead of the exception, which of course makes it easier for the next manager to "be innovative." Please note that repeatable processes also serve to reduce risk.

Capital One uses a model based on continuous innovation. It utilizes a patented Information-Based Strategy (IBS) that enables the company to expand its mature credit card business by tailoring more than 16,000 different product combinations to customers' needs. The company is able to embrace high degrees of risk because it bases its innovations on customer needs. The company tests new ideas against existing customers or possibly a separate grouping of prospects.

### Measuring Innovation

A wealth of metrics can be derived from the preceding discussions. Other innovation metrics to consider include:

- Return on innovation investment—Number of customers who view the brand as innovative divided by the total number of potential customers
- Brand innovation quotient—Number of repeat purchasers divided by total number of purchasers
- Pipeline process flow—Measures number of products at every stage of development (i.e., concept development, business analysis, prototype, test, launch)
- Innovation loyalty—Number of repeat purchases made before switching to a competitor

Citibank long had an Innovation Index. This index measured revenues derived from new products, but Citibank deemed this index insufficient to meet its needs. Citibank created an Innovation Initiative, staffed by a special task force. This group was challenged to come up with more meaningful metrics that could be used to track progress and be easily integrated into Citibank's balanced scorecard. The task force eventually developed a set of metrics, which included new revenue from innovation, successful transfer of products from one country or region to another, the number and type of ideas in the pipeline, and time from idea to profit.

A few years ago I wrote one or two books on using the Balanced Scorecard for performance management and measurement. The Balanced Scorecard views the business organization through four perspectives. Suffice it to say that there are several notable risk-related metrics that can be used within the Balanced Scorecard internal business processes perspective. More on Balanced Scorecard can be found in later in this book.

### The Six Steps to Increasing Creativity

Creativity stems from a contented workplace. Employees who feel fulfilled tend to give their all to the task at hand. In addition, reduced stress and those feelings of contentment actually promote reduced risk. The steps for increasing creativity follow.

*Step 1*

Make sure your company's goals are consistent with your value system. This is an interesting perspective in an era of few good jobs and trends toward outsourcing/offshoring and increasing automation/robotization of the workforce.

Many studies of employees at work find that job insecurity is a big issue; more permanent jobs are being replaced by temporary jobs; and employees are experiencing unrealistic expectations, increased workloads, and lack of staff. Unfortunately, few workers can imagine their situation improving substantially. In fact, many believe things will only get worse. They see the gap between rich and poor growing; and they feel that they do not know where all this change will end up.

Managers need to realize that many, if not most, employees will harbor some or all of these feelings. Given the dire job situation they might opt to stay put rather than find a company whose goals match their values. Therefore, it is up to the manager to somehow ameliorate the level of anxiety that accompanies these feelings such that creativity is not stifled.

*Step 2*

Pursue some self-initiated activity by choosing projects where your motivation is high. Few employees get to choose their own task assignments. However, savvy managers need to be aware that creativity is greatly enhanced when employees are motivated to do their jobs.

The better companies try to fit the employee to the task by creating a skills database. These permit managers to rapidly locate an employee who has the skills—and the motivation—to fulfill a particular work requirement. However, there will always be those times when the task the employee is expected to complete is simply not one he or she is especially interested in. At this point the good information technology manager will use a variety of motivating techniques.

The top five motivating techniques are:

1. Manager personally congratulates employee who does a good job.
2. Manager writes personal notes about good performance.
3. Organization uses performance as basis for promotion.
4. Manager publicly recognizes employee for good performance.
5. Manager holds morale-building meetings to celebrate successes.

One does not have to actually give an award for recognition to happen. Giving your attention is just as effective. The Hawthorne effect says the act of measuring (paying attention) will itself change behavior. There are some low-cost rewards recognition techniques that can be used to improve morale:

- Make a photo collage about a successful project that shows the people who worked on it, its stages of development, and its completion and presentation.
- Create a "yearbook" to be displayed in the lobby that contains each employee's photograph, along with his or her best achievement of the year.

- Establish a place to display memos, posters, photos, and so on, recognizing progress toward goals and thanking individual employees for their help.
- Develop a "Behind the Scenes Award" specifically for those whose actions are not usually in the limelight.
- Say thanks to your boss, your peers, and your employees when they have performed a task well or have done something to help you.
- Make a thank-you card by hand.
- Cover the person's desk with balloons.
- Bake a batch of chocolate-chip cookies for the person.
- Make and deliver a fruit basket to the person.
- Give a person a candle with a note saying, "No one holds a candle to you."
- Give a person a heart sticker with a note saying, "Thanks for caring."
- Purchase a plaque, stuffed animal, or anything fun or meaningful, and give it to an employee at a staff meeting with specific praise. That employee displays it for a while, then gives it to another employee at a staff meeting in recognition of an accomplishment.
- Call an employee into your office (or stop by his or her office) just to thank him or her; do not discuss any other issue.
- Post a thank-you note on the employee's office door.
- Send an e-mail thank-you card.
- Praise people immediately. Encourage them to do more of the same.
- Greet employees by name when you pass them in the hall.
- Make sure you give credit to the employee or group that came up with an idea being used.
- Acknowledge individual achievements by using employees' names when preparing status reports.
- Use social media to recognize employees.

It is important that you set up your employees for success. When you give someone a new assignment, tell the employee why you are trusting him or her with this new challenge. "I want you to handle this because I like the way you handled _____ last week." It is also important that you never steal the stage. When an employee tells you about an accomplishment do not steal her thunder by telling her about a similar accomplishment of yours. It is also important that you never use sarcasm, even in a teasing way. Resist the temptation to say something like, "It's about time you gave me this report on time." Deal with the "late" problem by setting a specific time the report is due. If it is done on time, make a positive comment about timeliness.

*Step 3*

Take advantage of unofficial activity. I know of few people who have the luxury of working on unofficial projects in larger companies. However, this is actually quite a good idea. Management should allow slack time to be used for creative purposes. Channels should be put in place such that any great idea nurtured during slack time has an equal opportunity to be presented for possible funding.

*Step 4*

Be open to serendipity. Scotchgard was actually invented by accident. As a manager, it is very important that I be open to this sort of novel product development.

*Step 5*

Diversify your stimuli. It is a good idea to rotate into every job the employee is capable of doing to induce intellectual cross-pollination. Rotating jobs is also a tenet of quality management systems, including ISO 9001.

*Step 6*

Create opportunities for information communication, otherwise known as "meet and greet." Salespeople are natural networkers. These folks sign up for every event and learn a great deal by doing so. Other employees are somewhat less motivated to leave the office to attend industry-wide gatherings, particularly as the employee gets older and has additional familial responsibilities.

The six steps for increasing creativity are but a starting point for creating the innovative organization. All of this, however, still relies on the CEO being an advocate for innovation management.

### Rewarding Employees for Innovative Ideas

Intrinsic rewards appeal to a person's desire for self-actualization, curiosity, joy, and interest in the work. Extrinsic rewards appeal to a person's desire for attainment (e.g., money, stock options, days offs, tickets to ballgames). Intrinsic rewards are intangible, whereas extrinsic rewards are quite tangible. As one of my employees says, "show me the money."

Many of the motivation techniques discussed here could be considered intrinsic rewards. Extrinsic reward systems are more difficult to implement, as there are usually budget considerations to deal with. In many companies, the methodology used to grant yearly raises can even be considered countermotivational. When I worked for the New York Stock Exchange, employees were rated on a scale of 1 to 5. The largest "rewards" (i.e., raises) were granted to the 5s. However, we were told to rate our employees using a bell-shaped curve. The result is that some 5s were cheated out of their fair share of the reward system.

Generating innovation can be assisted by using one or more bottom-up creativity techniques. A brief list of the best of these techniques follows:

- Brainstorming—This technique is perhaps the most familiar of all the techniques discussed here. It is used to generate a large quantity of ideas in a short period of time.
- Blue slip—Ideas are individually generated and recorded on a 3″ × 5″ sheet of blue paper. People anonymously share ideas in order to make people feel more at ease. Since each idea is on a separate piece of blue paper, the sorting and grouping of similar ideas is facilitated.
- Extrapolation—A technique or approach, already used by the organization, is stretched to apply to a new problem.
- Progressive abstraction technique—By moving through progressively higher levels of abstraction, it is possible to generate alternative problem definitions from an original problem.

When a problem is enlarged in a systematic way, it is possible to generate many new definitions that can then be evaluated for their usefulness and feasibility. Once an appropriate level of abstraction is reached, possible solutions are more easily identified.

- 5Ws and H technique—This technique is the traditional and journalistic approach of who, what, where, when, why, how. Use of this technique serves to expand a person's view of the problem and to assist in verifying that all related aspects of the problem have been addressed and considered.
- Force field analysis technique—The name of this technique comes from its ability to identify forces contributing to or hindering a solution to a problem. This technique stimulates creative thinking in three ways: (1) it defines direction, (2) identifies strengths that can be maximized, and (3) identifies weaknesses that can be minimized.
- Problem reversal—Reversing a problem statement often provides a different framework for analysis. For example, in attempting to come up with ways to improve productivity, try considering the opposite—how to decrease productivity.
- Associations/image technique—Most of us have played the game, at one time or another, that a person names a person, place, or thing and then asks for the first thing that pops into the second person's mind. The process of combining and linking is another way of expanding the solution space.
- Wishful thinking—This technique enables people to loosen analytical parameters to consider a larger set of alternatives than they might ordinarily consider. By permitting a degree of fantasy into the process, the result just might be a new and unique approach.

### Conclusion

Oren Harari (1993), a professor at the University of San Francisco and a management consultant, relates an interesting experience with one of his clients. While he was waiting for an appointment with this particular client, he overheard two of the manager's clerical assistants calling customers and asking them how they liked the company's product. Harari reflected that it was no wonder this manager had such a good reputation. When he finally met with her, he offered his congratulations on her ability to delegate the customer service task to her staff. "What you talking about?" she asked, bewildered. "Why, your secretaries are calling up customers on their own," Harari replied. "Oh, really? Is that what they're doing?" she laughed. "You mean you didn't delegate that task to them?" "No," she said. "I didn't even know they were doing it. Listen, Oren, my job is to get everyone on my team to think creatively in pursuit of the same goal. So what I do is talk to people regularly about why we exist as a company and as a team. That means we talk straight about our common purpose and the high standards we want to achieve. I call these our goal lines. Then we talk regularly about some broad constraints we have to work with them, like budgets, ethics, policies, and legalities. Those are our sidelines.

"It's like a sport. Once we agree on the goal lines and sidelines, I leave it to my people to figure out how to best get from here to there. I'm available and attentive when they need feedback. Sometimes I praise; sometimes I criticize—but always constructively, I hope. We get together periodically and talk about who's been trying what, and we give constructive feedback to one another. I know that sounds overly simplistic, but I assure you that is my basic management philosophy.

"And that's why I don't know what my assistants are doing, because it's obviously something they decided to try for the first time this week. I happen to think it's a great idea, because it's within the playing field and helps keep high standards for being number one in our industry. I will tell you something else: I don't even know what they intend to do with the data they're collecting, but I know they'll do the right thing.

"Here's my secret: I don't know what my people are doing, but because I work face to face with them as a coach, I know that whatever it is they're doing is exactly what I'd want them to be doing if I knew what they were doing!"

The Harari story is one of my favorites because it encapsulates into one very brief story exactly what it is a good manager is supposed to do to encourage innovative thinking in his or her employees. The key is to promote creativity and innovation while coaching toward reduced risk.

## Reference

Harari, O. 1993, Stop Empowering Your People. *Management Review*, November, pp. 26–29.

# 6

# PERFORMANCE MEASUREMENT AND MANAGEMENT FOR REDUCED RISK

Robert S. Kaplan and David P. Norton developed the balanced scorecard approach in the early 1990s to compensate for the perceived shortcomings of using only financial metrics to judge corporate performance. They recognized that in this new economy it was also necessary to value intangible assets. Because of this they urged companies to measure such esoteric factors as quality and customer satisfaction. By the middle 1990s, a balanced scorecard became the hallmark of a well-run company, one that saw reduced risk in the planning and implementation of its projects and processes. Kaplan and Norton often compare their approach for managing a company to that of pilots viewing assorted instrument panels in an airplane cockpit—both have a need to monitor multiple aspects of their working environment.

In the scorecard scenario, as shown in Figure 6.1, a company organizes its business goals into discrete, all-encompassing perspectives: financial, customer, internal process, and learning/growth. The company then determines cause-effect relationships, for example, satisfied customers buy more goods, which increases revenue. Next, the company lists measures for each goal, pinpoints targets, and identifies projects and other initiatives to help reach those targets.

Departments create scorecards tied to the company's targets, and employees and projects have scorecards tied to their department's targets. This cascading nature provides a line of sight between each individual, the project they're working on, the unit they support, and how that impacts the strategy of the enterprise as a whole.

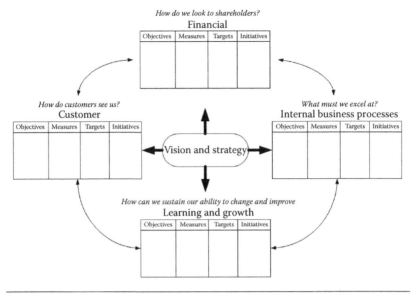

**Figure 6.1** The balanced scorecard.

For project managers, the balanced scorecard is an invaluable tool that permits the project manager to link a project to the business side of the organization using a cause-and-effect approach. Some have likened balanced scorecard to a new language, which enables the project manager and business line managers to think together about can be done to support and/or improve business performance.

A beneficial side effect of the use of the balanced scorecard is that when all measures are reported, one can calculate the strength of relations between the various value drivers. For example, if the relation between high implementation costs and high profit levels is weak for a long time, it can be inferred that the project, as implemented, does not sufficiently contribute to results as expressed by the other (e.g., financial, risk) performance measures.

### Adopting the Balanced Scorecard

Kaplan and Norton (2001) provide a good overview of how a typical company adapts to the balanced scorecard approach:

> Each organization we studied did it a different way, but you could see that, first, they all had strong leadership from the top. Second, they

translated their strategy into a balanced scorecard. Third, they cascaded the high-level strategy down to the operating business units and the support departments. Fourth, they were able to make strategy everybody's everyday job, and to reinforce that by setting up personal goals and objectives and then linking variable compensation to the achievement of those target objectives. Finally, they integrated the balanced scorecard into the organization's processes, built it into the planning and budgeting process, and developed new reporting frameworks as well as a new structure for the management meeting.

The key, then, is to develop a scorecard that naturally builds in cause-and-effect relationships, includes sufficient performance drivers and, finally, provides a linkage to appropriate measures, as shown in Table 6.1.

At the very lowest level, a discrete project can also be evaluated using a balanced scorecard. The key here is the connectivity between the project and the objectives of the organization as a whole, as shown in Table 6.2.

The internal processes perspective maps neatly to the traditional triple constraint of project management, using many of the same measures traditionally used. For example, we can articulate the quality constraint using the ISO 10006:2003 standard. This standard provides guidance on the application of quality management in projects. It is applicable to projects of varying complexity, small or large, of short or long duration, in different environments, and irrespective of the kind of product or process involved. Quality management of projects in this International Standard is based on eight quality management principles:

1. Customer focus
2. Leadership
3. Involvement of people
4. Process approach
5. System approach to management
6. Continual improvement
7. Factual approach to decision making
8. Mutually beneficial supplier relationships

Sample characteristics of these principles can be seen in Table 6.3.

**Table 6.1** Typical Departmental Sample Scorecard

| OBJECTIVE | MEASURE/METRICS | END OF FY 2010 (PROJECTED) |
|---|---|---|
| *Financial* | | |
| Long-term corporate profitability | % Change in stock price attributable to | +25% per year for next 10 years |
| | Earnings growth | +20% per year for next 10 years |
| Short-term corporate profitability<br>1. New products<br>2. Enhance existing products<br>3. Expand client-base<br>4. Improve efficiency and cost-effectiveness | Revenue growth<br>% Cost reduction | +20% related revenue growth<br>Cut departmental costs by 35% |
| *Customer* | | |
| Customer satisfaction<br>1. Customer-focused products<br>2. Improve response time<br>3. Improve security | Quarterly and annual customer surveys satisfaction index<br>Satisfaction ratio based on customer surveys | +35%; Raise satisfaction level from current 60% to 95%<br><br>+20% |
| Customer retention | % Of customer attrition | −7%; Reduce from current 12% to 5% |
| Customer acquisition | % Of increase in number of customers | +10% |
| *Internal* | | |
| Complete M&A transitional processes | % of work completed | 100% |
| Establish connectivity | % of workforce full access to corporate resources | 100% |
| Improve quality | % Saved on reduced work | +35% |
| Eliminate errors and system failures | % Reduction of customer complaints | +25% |
| | % Saved on better quality | +25% |
| Increase ROI | % Increase in ROI | +20%–40% |
| Reduce TCO | % Reduction of TCO | −10%–20% |
| Increase productivity | % Increase in customer orders | +25 |
| | % Increase in production/ employee | +15% |
| Product and services enhancements | Number of new products and services introduced | 5 new products |
| Improve response time | Average # of hours to respond to customer | −20 minutes; reduce from current level of 30–60 minutes to only 10 minutes or less |

(*Continued*)

**Table 6.1 (Continued)**  Typical Departmental Sample Scorecard

| OBJECTIVE | MEASURE/METRICS | END OF FY 2010 (PROJECTED) |
|---|---|---|
| *Learning and Innovations* | | |
| Development of skills | % Amount spent on training | +10% |
| Leadership development and training | % Staff with professional certificates | + 20 |
| | Number of staff attending colleges | 18 |
| Innovative products | % Increase in revenue | +20% |
| Improved process | Number of new products | +5 |
| R&D | % Decrease in failure, complaints | −10% |
| Performance measurement | % Increase in customer satisfaction survey results | +20 |
| | % Projects to pass ROI test | +25% |
| | % Staff receiving bonuses on performance enhancement | +25% |
| | % Increase in documentation | +20% |

**Table 6.2**  A Simple Project Scorecard Approach

| PERSPECTIVE | GOALS |
|---|---|
| Customer | Fulfill project requirements |
| | Control cost of the project |
| | Satisfying project end-users |
| | Collaborating with end-users leads to an improve set of requirements, which reduces risk.* |
| Financial | Provides business value (e.g., ROI, ROA, etc.) |
| | Project contributing to organization as a whole |
| | Reduction in costs due to enhanced communications.* |
| Internal processes | Adheres to triple constraint: time, cost, quality. |
| | Meeting the triple constraint leads to reduced risk.* |
| Learning and growth | Maintaining currency |
| | Anticipate changes |
| | Acquired skill sets |
| | Keeping systems current reduces risk.* |

*Possible goals related to risk.

Characteristics of a variable (e.g., quality, time) are used to create the key performance indicators (KPIs), or metrics, used to measure the "success" of the project. Thus, as you can see from Tables 6.1 to 6.3, we have got quite a few choices in terms of measuring the quality dimension of any particular project, as well as directly tie-in the social enterprising aspects of all of this. More specifically, the perspective

**Table 6.3**    ISO 10006 Definition of Quality Management for Projects

| QUALITY CHARACTERISTIC | SUBCHARACTERISTIC |
|---|---|
| Customer focus | 1. Understanding future customer needs<br>2. Meet or exceed customer requirements<br>Close collaboration with the various stakeholder groups lowers risk.* |
| Leadership | 1. By setting the quality policy and identifying the objectives (including the quality objectives) for the project<br>2. By empowering and motivating all project personnel to improve the project processes and product<br>Leaderful teams (where the teams lead themselves) reduces risk.* |
| Involvement of people | 1. Personnel in the project organization have well-defined responsibility and authority<br>2. Competent personnel are assigned to the project organization<br>The use of collaborative social and other technologies is naturally involving, which will hopefully lead to improved product and process, and serve to reduce risk.* |
| Process approach | 1. Appropriate processes are identified for the project<br>2. Interrelations and interactions among the processes are clearly identified<br>Properly configured environments enable the team to more effectively and quickly articulate business processes, and provides an excellent means for documenting those processes and articulated risk contingencies.* |
| System approach to management | 1. Clear division of responsibility and authority between the project organization and other relevant interested parties<br>2. Appropriate communication processes are defined<br>Employing a systematized method will lead to more effective management of the project, as well as enhanced communication among the various stakeholder groups. These attributes serve to lower risk.* |
| Continual improvement | 1. Projects should be treated as a process rather than as an isolated task<br>2. Provision should be made for self-assessments<br>Social enterprising provides the means for constant assessment via the group workspaces. This leads to reduced risk.* |
| Factual approach to decision making | 1. Effective decisions are based on the analysis of data and information<br>2. Information about the project's progress and performance are recorded<br>Tracking project progress and performance is critical to promoting a risk-aware project.* |
| Mutually beneficial supplier relationships | 1. The possibility of a number of projects using a common supplier is investigated<br>Working more collaboratively with suppliers reduces risk.* |

*Those characteristics with a tie-in to risk.

**Table 6.4**   Representative Risk Metrics

| INTERNAL BUSINESS PROCESS |
| --- |
| Definitional uncertainty risk |
| Technological risk |
| Development risk |
| Nonalignment risk |
| IT service delivery risk |
| Development risk |
| Maturity risk |

that best fits the risk planning and control paradigm is internal business processes. Possible metrics are shown in Table 6.4.

**Attributes of Successful Project Management Measurement Systems**

There are certain attributes that set apart successful performance measurement and management systems, including:

- *A conceptual framework is needed for the performance measurement and management system.* A clear and cohesive performance measurement framework that is understood by all managers and staff, and that supports objectives and the collection of results is needed.
- *Effective internal and external communications are the keys to successful performance measurement.* Effective communication with stakeholders is vital to the successful development and deployment of performance measurement and management systems.
- *Accountability for results must be clearly assigned and well-understood.* Managers must clearly identify what it takes to determine success and make sure that staff members understand what they are responsible for in achieving these goals.
- *Performance measurement systems must provide intelligence for decision makers, not just compile data.* Performance measures should relate to strategic goals and objectives, and provide timely, relevant, and concise information for use by decision makers—at all levels—to assess progress toward achieving predetermined goals. These measures should produce information on the efficiency with which resources (i.e., people,

hardware, software) are transformed into goods and services, on how well results compare to a program's intended purpose, and on the effectiveness of activities and operations in terms of their specific contribution to program objectives.

- *Compensation, rewards, and recognition should be linked to performance measurements.* Performance evaluations and rewards need to be tied to specific measures of success, by linking financial and nonfinancial incentives directly to performance. Such a linkage sends a clear and unambiguous message as to what is important.

- *Performance measurement systems should be positive, not punitive.* The most successful performance measurement systems are not "gotcha" systems, but *learning* systems that help identify what works—and what does not—so as to continue with and improve on what is working and repair or replace what is not working.

- *Results and progress toward program commitments should be openly shared with employees, customers, and stakeholders.* Performance measurement system information should be openly and widely shared with employees.

You will note that quite a few of these attributes seem to be made for social enterprising environments. Performance measurement systems should be communicated openly throughout the business. In all cases, however, for a balanced scorecard to work it has to be carefully planned and executed.

### Measuring Project Portfolio Management

Most organizations will have several ongoing programs all in play at once, all related to one or more business strategies. It is conceivable that hundreds of projects are ongoing, all in various stages of execution. Portfolio management is needed to provide the business and technical stewardship of all of these programs and their projects, as shown in Figure 6.2.

Portfolio management requires the organization to manage multiple projects at one time creating several thorny issues, the most salient ones are shown in Table 6.5. Many of the issues listed in Table 6.5

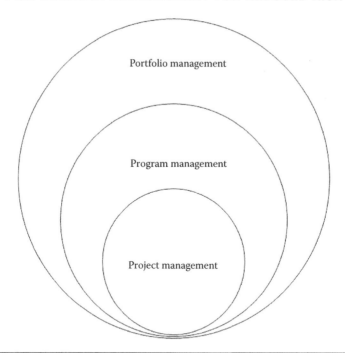

**Figure 6.2**  Portfolio management.

**Table 6.5**  Multiple Project Management Issues

| RESPONSIBILITY | ISSUE |
| --- | --- |
| Alignment management | Balancing individual project objectives with the organization's objectives |
| Control and communication | Maintaining effective communications within a project and across multiple projects |
| | Maintaining motivation across project teams |
| | Resource allocation |
| Learning and knowledge management | Inability to learn from past projects |
| | Failure to record lessons learned for each project |
| | Lack of timely information |

can be resolved using social enterprising techniques. Inter- and intra-project communications would be quite possible, as would maintaining motivation across project teams. Maintaining all of the project documentation online means that it would now be possible to record lessons learned. Thus, the knowledge gleaned during past projects would no longer be lost. Finally, information would be able to move through the system quickly and reach team members without delay or loss.

Portfolio management is usually performed by a project management office (PMO). This is the department or group that defines and maintains the standards of process within the organization. The PMO strives to standardize and introduce economies of repetition in the execution of projects. The PMO is the source of documentation, guidance, and metrics on the practice of project management and execution. Although most PMOs are independent of the various project teams, it might be worthwhile to assign to the PMO oversight of the social enterprising effort to ensure that there is some degree of standardization in terms of usage throughout the company.

A good PMO will base project management principles on accepted industry standard methodologies. Increasingly, influential industry certification programs such as ISO 9000 and the Malcolm Baldrige National Quality Award, government regulatory requirements such as Sarbanes-Oxley, and business process management techniques such as the balanced scorecard have propelled organizations to standardize processes.

If companies manage projects from an investment perspective—with a continuing focus on value, risk, cost, and benefits—costs should be reduced with an attendant increase in value. This is the driving principle of portfolio management.

A major emphasis of PMO is standardization. To achieve this end, the PMO employs robust measurement systems. For example, the following metrics might be reported to provide an indicator of process responsiveness:

1. Total number of project requests submitted, approved, deferred, and rejected.
2. Total number of project requests approved by the portfolio management group through the first project request approval cycle (this will provide an indicator of quality of project requests)
3. Total number of project requests and profiles approved by the portfolio management group through secondary and tertiary prioritization approval cycles (to provide a baseline of effort versus return on investment [ROI] for detailed project planning time)
4. Time and cost through the process

5. Changes to the project allocation after portfolio rebalancing (total projects, projects canceled, project postponed, projects approved)
6. Utilization of resources: percentage utilization per staff resource (over 100%, 80% to 100%, under 80%, projects understaffed, staff-related risks)
7. Projects canceled after initiation (project performance, reduced portfolio funding, reduced priority, and increased risk)

Interestingly, PMOs are not all that pervasive in industry. However, they are recommended if the organization is serious about enhancing performance and standardizing performance measurement. Implementation of a PMO is a project unto itself, consisting of three steps: take inventory, analyze, and manage.

1. A complete inventory of all initiatives should be developed. Information such as the project's sponsors and champion, stakeholder list, strategic alignment with corporate objectives, estimated costs, and project benefits should be collected.
2. Once the inventory is completed and validated, all projects on the list should be analyzed. A steering committee should be formed that has enough insight into the organization's strategic goals and priorities to place projects in the overall strategic landscape. The output of the analysis step is a prioritized project list. The order of prioritization is based on criteria that the steering committee selects. This is different for different organizations. Some companies might consider strategic alignment to be the most important, while other companies might decide that cost–benefit ratio is the better criterion for prioritization.
3. Portfolio management is not a one-time event. It is a constant process that must be managed. Projects must be continually evaluated based on changing priorities and market conditions.

It is the analyze step where the balanced scorecard should be created. The scorecard should be fine-tuned during the priotization of the project list outputted during the analysis step.

**Table 6.6**   Sample Risk-Related Metrics

| PROCESS | SUBPROCESS | ASSOCIATED SAMPLE METRIC |
|---|---|---|
| Initiating a Project (IP) | IP1 Planning Quality | Number of collaborative planning sessions promoting quality |
| | IP2 Planning a Project | % Resources devoted to risk planning |
| | IP3 Refining the Business Case and Risks | % Collaborative sessions where risk was assessed |

In all likelihood the PMO will standardize on a particular project management methodology. There are two major project management methodologies. The Project Management Body of Knowledge (PMBOK), which is most popular in the United States, recognizes five basic process groups typical of almost all projects: initiating, planning, executing, controlling and monitoring, and closing. Projects in Controlled Environments, PRINCE2, which is the de facto standard for project management in the United Kingdom and is popular in more than 50 other countries, defines a wide variety of subprocesses, but organizes these into eight major processes: starting a project, planning, initiating a project, directing a project, controlling a stage, managing product delivery, managing stage boundaries, and closing a project. Both PRINCE2 and PMBOK consist of a set of processes and associated subprocesses. These can be used to craft relevant risk-reducing metrics, as shown in Table 6.6.

Since the PMO is the single focal point for all things related to project management, it is natural that the project management balanced scorecard should be within the purview of this department.

### Project Management Process Maturity Model (PM)² and Collaboration

The PM² model determines and positions an organization's relative project management level with other organizations. There are a variety of project management process maturity models, and they are all based on work done by the Software Engineering Institute at Carnegie Mellon on improving the quality of the software development process.

The PM² model defines five steps, as shown in Figure 6.3. Unfortunately, quite a good number of organizations are still hovering somewhere between the ad-hoc and planned levels. Companies

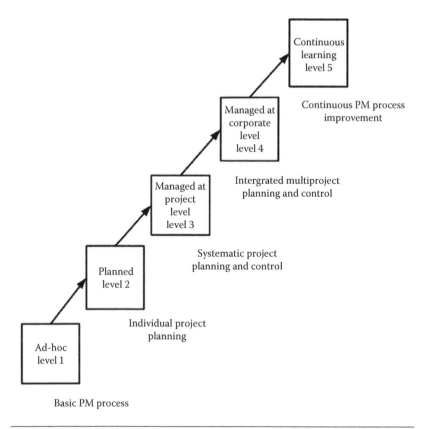

**Figure 6.3**  The PM² model.

that are serious about improving performance strive to achieve level 5, continuous learning. To do this requires a company to compare itself to others in its peer grouping, the goal of a model such as PM².

In the PM² model key processes, organizational characteristics and key focus areas are defined, as shown in Table 6.7. Each maturity level is associated with a set of key project management processes, characteristics of those processes, and key areas on which to focus. When mapped to the four balanced scorecard perspectives, PM² becomes a reference point or yardstick for best practices and processes.

Thus, measurement across collaborative, distributed partners must be considered in any measurement program. Several interest groups and partnerships in the automotive industry were formed to develop new project management methods and processes that worked effectively in a collaborative environment. The German Organization for Project Management (GPM e.V.), the PMI automotive special interest

**Table 6.7** Key Components of the PM² Model

| MATURITY LEVEL | KEY PM PROCESSES | MAJOR ORGANIZATIONAL CHARACTERISTICS | KEY FOCUS AREAS |
|---|---|---|---|
| Level 5 (Continuous Learning) | PM processes are continuously improved<br>PM processes are fully understood<br>PM data are optimized and sustained | Project-driven organization<br>Dynamic, energetic, and fluid organization<br>Continuous improvement of PM processes and practices | Innovative ideas to improve PM processes and practices |
| Level 4 (Managed at Corporate Level) | Multiple PM (program management)<br>PM data and processes are integrated<br>PM data are quantitatively analyzed, measured, and stored | Strong teamwork<br>Formal PM training for project team | Planning and controlling multiple projects in a professional manner |
| Level 3 (Managed at Project Level) | Formal project planning and control systems are managed<br>Formal PM data are managed | Team oriented (medium)<br>Informal training of PM skills and practices | Systematic and structured project planning and control for individual project |
| Level 2 (Planned) | Informal PM processes are defined<br>Informal PM problems are identified<br>Informal PM data are collected | Team oriented (weak)<br>Organizations possess strengths in doing similar work | Individual project planning |
| Level 1 (Ad-hoc) | No PM processes or practices are consistently available<br>No PM data are consistently collected or analyzed | Functionally isolated<br>Lack of senior management support<br>Project success depends on individual efforts | Understand and establish basic PM processes |

group, the automotive industry action group (AIAG), and others have embarked on projects to develop methods, models, and frameworks for collaborative product development, data exchange, quality standards, and project management. One recent output from this effort was the ProSTEP-iViP reference model to manage time, tasks, and communications in cross-company automotive product development projects (http://www.prostep.org/en/).

A set of drivers and KPIs for a typical stand-alone project can be seen in Table 6.8. Using guidelines from ProSTEP reference model, Niebecker, Eager, and Kubitza (2008) have reoriented the drivers and KPIs in Table 6.8 to account for the extra levels of complexity found in a project worked on by two or more companies in a networked collaborative environment. This suits the risk planning construct quite nicely, as shown in Table 6.9.

**Table 6.8**   Representative Drivers and KPIs for a Standard Project

| BALANCED SCORECARD PERSPECTIVE | DRIVERS | KPIs |
| --- | --- | --- |
| Finances | Project budget<br>Increase of business value<br>Multiproject categorization<br>Project management | Human resources<br>Share of sales<br>Profit margin<br>Savings<br>ROI<br>Expenditure |
| Customer | Customer satisfaction | Cost overrun<br>Number of customer audits<br>Change management<br>Process stability |
| Process | Adherence to schedules<br>Innovation enhancement<br>Minimizing risks<br>Optimization of project<br>  structure<br>Quality | Adherence to delivery dates<br>Lessons learned<br>Number of patent applications<br>External labor<br>Quality indices<br>Duration of change management<br>Product maturity<br>Percentage of overhead<br>Number of internal audits<br>Project risk analysis |
| Development | Employee satisfaction<br>Employee qualification<br>  enhancement | Rate of employee fluctuation<br>Travel costs<br>Overtime<br>Index of professional experience<br>Continuing education costs |

**Table 6.9**    Drivers and KPIs for a Collaborative Project (CP)

| BALANCED SCORECARD PERSPECTIVE | DRIVERS | KPIs |
|---|---|---|
| Finances/Project | Project cost | Product costs |
| | Increase of business value | Production costs |
| | Categorization into CP | Cost overruns |
| |   management | Savings |
| | Project maturity | Productivity index |
| | | Turnover |
| | | Risk distribution |
| | | Profit margin |
| | | Feature stability |
| | | Product maturity index |
| Process | Adherence to schedules | Variance to schedule |
| | Innovation enhancement | Changes before and after design |
| | Minimizing risks |   freeze |
| | Adherence to collaboration | Duration until defects removed |
| |   process | # and duration of product changes |
| | Quality | # of postprocessing changes |
| | | Continuous improvement process |
| | | Project risk analysis |
| | | Maturity of collaboration process |
| | | Frequency of product tests |
| | | Defect frequency |
| | | Quality indices |
| Collaboration | Communication | Number of team workshops |
| | Collaboration | Checklists |
| | | Degree of communication efficiency |
| | | Collaborative lessons learned |
| | | Maturity of collaboration |
| | | Degree of lessons learned realization |
| Development | Team satisfaction | Employee fluctuation |
| | Team qualification | Project focused continuing education |
| |   enhancement | Employee qualification |
| | Trust between team | |
| |   members | |

Niebecker, Eager, and Kubitza's recommendations expand upon the traditional balanced scorecard methodology, providing an approach for monitoring and controlling cross-company projects by aligning collaborative project objectives with the business strategies and project portfolio of each company.

## Conclusion

In this chapter we presented a strong case for using performance management and measurement techniques to reduce the risks of planning and implementing projects and process. In the next chapter we will incorporate all of this in a discussion on information technology project tracking.

# References

Kaplan, R. S., and D. P. Norton. 2001. On Balance (interview). *CFO, Magazine for Senior Financial Executives*, February 1.

Niebecker, K., D. Eager, and K. Kubitza. 2008. Improving cross-company management performance with a collaborative project scorecard. *International Journal of Managing Projects in Business* 1(3):368–386.

# 7

# INFORMATION TECHNOLOGY PROJECT TRACKING AND CONTROL

In Chapter 6 we dived deep into a discussion on performance management and measurement, focusing on the Balanced Scorecard technique. We will incorporate those techniques in this chapter as we cover the concepts of project tracking and control. Emphasis will be placed on status reporting, project metrics, and process improvement, all of which serve to manage risk.

## Things Change

The project plan is all-encompassing. It includes a full detailing of estimates for project costs, resources, and scheduling. Essentially, it forms the baseline for the management of the implementation of the project. However, both internal as well as external environmental risk factors do impact the cost, schedule, and even viability of the project. These environmental factors include:

1. Requirements change. It is not uncommon for end-users to start crafting their wish lists before a system ever gets placed in production. For the most part, project managers freeze changes to a system while the system is being constructed, hence the wish list. Most of the time, changed requirements can be accommodated by the wish list process. However, there are occasions when the requirements have changed so dramatically there is no choice but to terminate the system and, perhaps, start again.

2. Regulations change. Many organizations are governed by a variety of state, local, and federal regulatory guidelines. These have a habit of changing. For example, the cellular phone number portability mandate will require extensive change to

many systems. There might well be one or two in development mode that might be terminated altogether as a result of this regulatory modification.

3. New technologies are introduced.
4. Technologies in use are deemed obsolete.
5. Functionality is outsourced and/or merged into another department or company. Mergers and acquisitions have a way of terminating projects midstream for obvious reasons. Today's organizational propensity for outsourcing jobs to lower-cost countries impact a project.
6. Employee illness, vacations, departure, and unanticipated developer-related difficulties hinder developing the system.

### Budget Monitoring

The budget approved by management (i.e., the estimated budget for the time cycle) is referred to as the baseline budget. The key word here is *base*, meaning that which actual budget costs are compared to.

Project budgets are usually monitored on a continuous basis with a status report submitted to management on a monthly basis. The status report will report on any variations (i.e., actual budget) from this baseline budget as well as projections (i.e., projected budget) for future status reporting periods.

The goal of budget monitoring is to keep project costs within an acceptable deviation (e.g., 10%) from the baseline. There are several reasons why a budget would deviate from the baseline:

- Estimation error
- Priority change
- Resource cost change (e.g., contractor quits and a new one is hired)
- Additional requirements added to project

### Project Reporting

Project implementation is a dynamic rather than static process. Because it is dynamic, the project manager needs to carefully track the project to ensure that it is completed on time and within budget.

Figure 7.1 shows a typical baseline schedule for a project as developed using Microsoft's Project software tool. Each software developer is usually responsible for filing a weekly report detailing progress of his or her assigned tasks. The project manager uses these reports to alter the baseline, if necessary, resulting in a report such as the one shown in Figure 7.2. In the same way, a project's costs can be managed, as shown in Figure 7.3, using automated tools. Both the baseline and variances to the baseline can be tracked.

**Figure 7.1**   Using Microsoft Project to manage a project's schedule.

| ID | Task Name | Work | Baseline | Variance | Actual | Remaining | % W. Comp. |
|----|-----------|------|----------|----------|--------|-----------|------------|
| 1 | Develop Feature Requirements Statement | 196 hrs | 196 hrs | 0 hrs | 0 hrs | 196 hrs | 0% |
| 2 | Complete Market Research | 42 hrs | 42 hrs | 0 hrs | 0 hrs | 42 hrs | 0% |
| 3 | Develop Management Business Case | 30 hrs | 30 hrs | 0 hrs | 0 hrs | 30 hrs | 0% |
| 4 | Present Business Case for Approval | 4 hrs | 4 hrs | 0 hrs | 0 hrs | 4 hrs | 0% |
| 5 | Management Approval of Business Case | 3.6 hrs | 3.6 hrs | 0 hrs | 0 hrs | 3.6 hrs | 0% |
| 6 | Develop Feature Design Document | 400 hrs | 400 hrs | 0 hrs | 0 hrs | 400 hrs | 0% |
| 7 | Complete Design Review | 126 hrs | 126 hrs | 0 hrs | 0 hrs | 126 hrs | 0% |
| 8 | Complete Design Approval | 8 hrs | 8 hrs | 0 hrs | 0 hrs | 8 hrs | 0% |
| 9 | Develop Technical Specifications | 56 hrs | 56 hrs | 0 hrs | 0 hrs | 56 hrs | 0% |
| 10 | Complete Approval of Technical Specifications | 16 hrs | 16 hrs | 0 hrs | 0 hrs | 16 hrs | 0% |
| 11 | Develop Server Code | 1,260 hrs | 1,260 hrs | 0 hrs | 0 hrs | 1,260 hrs | 0% |
| 12 | Develop GUI Code | 1,260 hrs | 1,260 hrs | 0 hrs | 0 hrs | 1,260 hrs | 0% |
| 13 | System Testing | 320 hrs | 320 hrs | 0 hrs | 0 hrs | 320 hrs | 0% |
| 14 | Integration Testing | 168 hrs | 168 hrs | 0 hrs | 0 hrs | 168 hrs | 0% |
| 15 | Regression Testing | 120 hrs | 120 hrs | 0 hrs | 0 hrs | 120 hrs | 0% |
| 16 | Feature Documentation Developed | 680 hrs | 680 hrs | 0 hrs | 0 hrs | 680 hrs | 0% |
| 17 | Operational Packaging | 50 hrs | 50 hrs | 0 hrs | 0 hrs | 50 hrs | 0% |
| 18 | QA Acceptance Testing | 96 hrs | 96 hrs | 0 hrs | 0 hrs | 96 hrs | 0% |
| 19 | QA Acceptance | 96 hrs | 96 hrs | 0 hrs | 0 hrs | 96 hrs | 0% |

**Figure 7.2**   Tracking variances to the baseline.

| ID | Task Name | Fixed Cost | Fixed Cost Accrual | Total Cost | Baseline | Variance | Actual | Remaining |
|----|-----------|-----------|--------------------|-----------|----------|----------|--------|-----------|
| 1 | Develop Feature Requirements Statement | $0.00 | Prorated | $12,700.00 | $12,700.00 | $0.00 | $0.00 | $12,700.00 |
| 2 | Complete Market Research | $10,000.00 | Prorated | $13,500.00 | $13,500.00 | $0.00 | $0.00 | $13,500.00 |
| 3 | Develop Management Business Case | $0.00 | Prorated | $2,350.00 | $2,350.00 | $0.00 | $0.00 | $2,350.00 |
| 4 | Present Business Case for Approval | $0.00 | Prorated | $400.00 | $400.00 | $0.00 | $0.00 | $400.00 |
| 5 | Management Approval of Business Case | $0.00 | Prorated | $320.00 | $320.00 | $0.00 | $0.00 | $320.00 |
| 6 | Develop Feature Design Document | $0.00 | Prorated | $22,333.33 | $22,333.33 | $0.00 | $0.00 | $22,333.33 |
| 7 | Complete Design Review | $0.00 | Prorated | $7,342.78 | $7,342.78 | $0.00 | $0.00 | $7,342.78 |
| 8 | Complete Design Approval | $0.00 | Prorated | $700.00 | $700.00 | $0.00 | $0.00 | $700.00 |
| 9 | Develop Technical Specifications | $0.00 | Prorated | $3,120.00 | $3,120.00 | $0.00 | $0.00 | $3,120.00 |
| 10 | Complete Approval of Technical Specifications | $0.00 | Prorated | $1,400.00 | $1,400.00 | $0.00 | $0.00 | $1,400.00 |
| 11 | Develop Server Code | $15,500.00 | Prorated | $88,862.50 | $88,862.50 | $0.00 | $0.00 | $88,862.50 |
| 12 | Develop GUI Code | $6,000.00 | Prorated | $78,862.50 | $78,862.50 | $0.00 | $0.00 | $78,862.50 |
| 13 | System Testing | $2,500.00 | Prorated | $18,900.00 | $18,900.00 | $0.00 | $0.00 | $18,900.00 |
| 14 | Integration Testing | $0.00 | Prorated | $6,960.00 | $6,960.00 | $0.00 | $0.00 | $6,960.00 |
| 15 | Regression Testing | $0.00 | Prorated | $6,400.00 | $6,400.00 | $0.00 | $0.00 | $6,400.00 |
| 16 | Feature Documentation Developed | $0.00 | Prorated | $58,400.00 | $58,400.00 | $0.00 | $0.00 | $58,400.00 |
| 17 | Operational Packaging | $5,000.00 | Prorated | $7,000.00 | $7,000.00 | $0.00 | $0.00 | $7,000.00 |
| 18 | QA Acceptance Testing | $0.00 | Prorated | $5,120.00 | $5,120.00 | $0.00 | $0.00 | $5,120.00 |
| 19 | QA Acceptance | $0.00 | Prorated | $9,320.00 | $9,320.00 | $0.00 | $0.00 | $9,320.00 |

**Figure 7.3**   Tracking project costs.

Project managers typically submit project status reports to end-users as well as senior management. Reports run the gamut from executive summaries to full-blown reports, which include outputs from tools such as Microsoft Project. In some organizations, monthly formal presentations are also required. In all cases, the project manager needs to learn the art of effective project communications to keep the project running smoothly as well as funded.

**Project Metrics**

As the project is progressing, the project manager should be collecting information, which will ultimately lead to a variety of business and technology metrics. The metrics that are reported upon are particular to the requirements of the level of management being reported to.

*Business Metrics*

There are a wide variety of business metrics that could be reported. Some of the most popular are shown in Figure 7.4.

*Technology Metrics*

Just as there is an abundance of business metrics, there is also an abundance of technology metrics. Some of the most common include:

- Lines of code
- Pages of documentation
- Number and size of tests
- Function count
- Variable count
- Number of modules
- Depth of nesting
- Count of changes required
- Count of discovered defects
- Count of changed lines of code
- Time to design, code, test
- Defect discovery rate by phase of development
- Cost to develop
- Number of external interfaces

| Category | Metric |
| --- | --- |
| Cost | • Actual cost vs. budget (variance) |
| | • Total cost per transaction |
| | • Labor costs vs. non-labor costs |
| | • Staff costs vs. consultant costs |
| Duration | • Actual time expended vs. budget (variance) |
| Productivity | • Effort hours per unit of work |
| | • Effort hours reduced from standard project processes |
| | • Effort hours saved through reuse |
| | • Number of process improvement ideas implemented |
| | • Number of hours/dollars saved from process improvements |
| Quality of deliverables | • Percentage of deliverables going through quality reviews |
| | • Percentage of deliverable reviews resulting in acceptance the first time |
| | • Number of defects discovered after initial acceptance |
| | • Percentage of deliverables that comply with organization standards |
| | • Number of hours to rework previously completed deliverables |
| | • Number of best practices identified and applied on the project |
| Client satisfaction | • Overall client satisfaction |
| | • Number of approved business requirements satisfied by the project |

**Figure 7.4** Popular business metrics.

- Number of tools used and why
- Reusability percentage
- Variance of schedule
- Staff years experience with team
- Staff years experience with language
- Software years experience with software tools
- Million instructions per second per person
- Support-to-development-personnel ratio
- Nonproject-to-project time ratio

Technology metrics measure both productivity as well as quality. The following checklist measures a variety of quality attributes:

Rate 1 to 5 (1 being the worst, 5 being the best)
- How easy is it to use?
- How secure is it?
- Level of confidence in it?
- How well does it conform to requirements?
- How easy is it to upgrade?
- How easy is it to change?
- How portable is it?
- How easy is it to locate a problem and fix?
- Is the response time fast enough?
- How easy is it to train staff?
- Ease of testing?
- Ease of coupling this system to another?
- Does the system utilize the minimum storage possible?
- Is the system self-descriptive?
- Is there a program for ongoing quality awareness for all employees?
- Is this the right system to be developed?

Other technology metrics include:

- Function points—Function points are a measure of the size of computer applications and the projects that build them. The size is measured from a functional, or user, point of view. It is independent of the computer language, development methodology, technology, or capability of the project team used to develop the application.
- Lines of code (LOC) metric—One of the common bases on which to estimate a software project is the lines of code (LOC) metric. LOC are used to determine time and cost estimates. The LOC estimate becomes the baseline to measure the degree of work performed on a project. Once a project is underway, the LOC becomes a tracking tool that can measure the degree of progress on a project. Experienced developers can gage a LOC estimate using prior knowledge of previous projects.

- Effective lines of code—An effective line of code is the measurement of all lines that are not comments on blank lines or standalone braces or parentheses. This measurement more closely represents the quantity of work performed. It is common for programmers to use a single brace or parenthesis on a line to denote a specific block of code. A single character on a line should not really count as a line of code. This type of coding style can therefore increase the LOC metric by 20 to 40 percent.

- Comment line metric—The amount of comments in a source program is a measure of the care taken by the programmer to make the source code and algorithms understandable. Code that is not well commented is very difficult to maintain. Comments can occur by themselves on a physical line or be comingled with source code. A line is considered a comment line if the physical line contains a comment.

- Blank line/white space metric—The number of blank lines within a program determines the readability. White space highlights the logical grouping of constructs and variables. Programs that use few blank lines are difficult to read and more expensive to maintain.

- Function count metric—The total number of functions within a program determines the degree of modularity. This metric is used to quantify the average number of LOC per function, maximum LOC per function, and the minimum LOC per function.

- Average LOC/function metric—The average LOC/function indicates how the code meets the accepted standard. The accepted industry standard of 200 LOC/function is desired as the average. Functions that have a larger number of lines of code per function are difficult to understand and difficult to maintain. They provide a good indication that a function should be broken into smaller functions.

- Maximum LOC/function metric—Although the average LOC per function gives an interesting source code trend, the maximum LOC per function gives an indication of the largest function in the system.

- Minimum LOC/function metric—A LOC value of 2 may indicate that functions are just prototypes and will need to be completed later.

*Risk Metrics*

Risk metrics depend on the type of system being built, but some risks are common to all systems:

- Percent of definitional uncertainty risk—Describes a low degree of project specification. Rate risk probability from 0% to 100%. Benchmark is ≤10%.
- Percent of technological risk—Assesses the use of bleeding edge technology. Rate risk probability from 0% to 100%. Benchmark is ≤45%.
- Percent of developmental risk—Tracks the lack of development skill sets among project staff. Benchmark is ≤10%.
- Percent of nonalignment risk—Assesses the resistance of employees and/or end-users to change. Rate probability of risk from 0% to 100%. Benchmark is ≤4%.
- Percent of IT service delivery risk—Assesses the problems with the delivering system (e.g., interface difficulties). Rate risk probability from 0% to 100%. Benchmark is ≤5%.
- Number of fraudulent transactions—Tracks the number of fraudulent transactions uncovered by analysis. Benchmark is ≤1%.
- Percent of systems that have risk contingency plans—Tracks the ratio of systems with contingency plans with those without. Benchmark is >95%.
- Percent of systems that have been assessed for security breaches—Tracks the ratio of systems that have been assessed for security breaches. Benchmark is ≥95%.

## Methods for Assessment

It is necessary to establish targets and means for assessment. The following procedure is not focused on any particular set of metrics. Rather it stresses that metrics should be selected on the basis of goals. This procedure is suitable for setting up goals for either the entire project deliverables or for any partial product created in the software life cycle.

1. Define measurable goals—The project goals establishment process is similar to the development process for project deliverables. Software projects usually start with abstract problem

concepts, and the final project deliverables are obtained by continuously partitioning and refining the problem into tangible and manageable pieces. Final quantified goals can be transformed from initial intangible goals by following the same divide-and-conquer method for software deliverables. Three sources of information are helpful to establishing the targets:

- Historical data under the assumptions that data are available, development environment is stable, and projects are similar in terms of type, size, and complexity.
- Synthetic data such as modeling results are useful if models used are calibrated to a specific development environment.
- Expert opinions.

2. Maintain balanced goals—The measurable goals are usually established on the basis of the following four factors: cost, schedule, effort, and quality. It is feasible to achieve just a single goal, but it is always a challenge to deliver a project with the minimum staff and resource, on time and within budget. It needs to be kept in mind that a trade-off is always involved and all issues should be addressed to reach a set of balanced goals.

3. Set up intermediate goals—A project should never be measured only at its end point. Checkpoints should be set up to provide confidence that the project is running on course. The common practice involves setting up quantifiable targets for each phase, measuring the actual values against the targets, and establishing a plan to make corrections for any deviations. All four aforementioned factors should be broken down into phase or activity for setting up intermediate targets. Measurements for cost and effort can be divided into machine and human resources according to the software life cycle phase so that expenditures can be monitored to ensure project is running within budget. The schedule should always be defined in terms of milestones or checkpoints to ensure intermediate products can be evaluated and the final product will be delivered on time. The quality of intermediate products should always be measured to guarantee the final deliverable will meet its target goal.

4. Establish means of assessment—Two aspects are involved in this activity: data collection and data analysis. Based on project characteristics such as size, complexity, and level of control, a decision should be made in terms of whether a manual data collection process or an automated data collection process should be used. If a non-automated way is applied, then the availability of the collection medium at the right time should be emphasized. For data analysis, the following two types of analyses should be considered:

- Project analysis—This type of analysis consists of checkpoint analysis and continuous analysis (trend analysis), and is concerned with verifying that intermediate targets are met to ensure that the project is on the right track.
- Component analysis—This type of analysis concentrates on the finer level of details of the end product, and is concerned with identifying those components in the product that may require special attention and action. The complete process includes deciding on the set of measures to be analyzed, identifying the components detected as anomalous using measured data, finding the root cause of the anomalies, and taking actions to make correction.

There are certain attributes that set apart successful performance measurement and management systems, including:

1. A conceptual framework is needed for the performance measurement and management system. Every organization, regardless of type, needs a clear and cohesive performance measurement framework that is understood by all levels of the organization, and that supports objectives and the collection of results.
2. Effective internal and external communications are the keys to successful performance measurement. Effective communication with employees, process owners, customers, and stakeholders is vital to the successful development and deployment of performance measurement and management systems.
3. Accountability for results must be clearly assigned and well understood. High-performance organizations clearly identify what it takes to determine success and make sure that all

managers and employees understand what they are responsible for in achieving organizational goals.

4. Performance measurement systems must provide intelligence for decision makers, not just compile data. Performance measures should be limited to those that relate to strategic organizational goals and objectives, and that provide timely, relevant, and concise information for use by decision makers—at all levels—to assess progress toward achieving predetermined goals. These measures should produce information on the efficiency with which resources are transformed into goods and services, on how well results compare to a program's intended purpose, and on the effectiveness of organizational activities and operations in terms of their specific contribution to program objectives.

5. Compensation, rewards, and recognition should be linked to performance measurements. Performance evaluations and rewards need to be tied to specific measures of success, by linking financial and nonfinancial incentives directly to performance. Such a linkage sends a clear and unambiguous message to the organization as to what is important.

6. Performance measurement systems should be positive, not punitive. The most successful performance measurement systems are not "gotcha" systems, but learning systems that help the organization identify what works—and what does not—so as to continue with and improve on what is working and repair or replace what is not working.

7. Results and progress toward program commitments should be openly shared with employees, customers, and stakeholders. Performance measurement system information should be openly and widely shared with an organization's employees, customers, stakeholders, vendors, and suppliers.

### Process Improvement

Modifying a system or process to enhance productivity or for competitive advantage is referred to as process improvement. Instituting a program that continually searches for ways to improve processes is referred to as continuous improvement.

Once a company decides that continuous improvement is indeed key to competitive advantage, it must create and then manage a continuous improvement plan. The next question that needs to be asked is whether the company is able to articulate a continuous improvement plan for its operations.

The goals of an information technology (IT) balanced scorecard are simplistic in scope but complex to execute:

1. Align IT plans with business goals and needs
2. Establish appropriate measures for evaluating the effectiveness of IT
3. Align employees' efforts toward achieving IT objectives
4. Stimulate and improve IT performance
5. Achieve balanced results across stakeholder groups

The keyword here is *balanced*. It reflects the balance between the aforementioned five goals listed, the four balanced scorecard perspectives (customer, business processes, learning and innovation, and financial), long- and short-term objectives, as well as between qualitative and quantitative performance measures.

Progressive scorecard practitioners track metrics in five key categories:

1. Financial performance—IT spending in the context of service levels, project progress, and so on. Sample metrics include cost of data communications per seat and relative spending per portfolio category.
2. Project performance—Sample metrics include percentage of new development investment resulting in new revenue streams and percentage of IT R&D investment leading to IT service improvements.
3. Operational performance—Instead of concentrating measurement efforts on day-to-day measures, best-in-class practitioners seek to provide an aggregate, customer-focused view of IT operations. Sample metrics include peak time availability and critical process uptime.
4. Talent management—This category of metrics seeks to manage IT human capital. Measures include staff satisfaction and retention as well as attractiveness of the IT department to

external job seekers. Metrics include retention of high-potential staff and external citations of IT achievement.

5. User satisfaction—Sample metrics include focused executive feedbacks and user perspective.

Bowne & Co., a New York City-based documents management company, has initiated an IT balanced scorecard. Its process consisted of seven steps:

1. Kickoff training for IT staff.
2. Ongoing strategy mapping—The annual IT strategy, like most companies, is derived from the corporate strategy.
3. Metrics selection—A team, including the chief technology officer, created a list of metrics. The list was refined using analysis of each potential metric's strengths and weaknesses. The final list was approved by the chief information officer (CIO).
4. Metrics definition—A set of standard definitions is created for each metric. It defines the measurement technique as well as the data collection process. It outlines initiatives that must be completed to allow tracking of the metrics.
5. Assigning metric ownership—Owners are assigned to each metric and they are responsible for scorecard completion. Their bonuses are related to their scorecard-related duties.
6. Data collection and quality assurance—Data frequency varies by metrics based on cost of collection, the corporate financial reporting cycle, and volatility of the business climate.
7. CIO, CTO, and corporate officers review the scorecard every six months. Metrics are revisited annually.

A sample balanced scorecard is shown in Figure 7.5.

Quality and productivity have always been an explicit part of Cupertino, California-based Hewlett-Packard's (HP) corporate objectives. To help develop and utilize metrics company-wide, HP created the Software Metrics Council. Today, eighty productivity and quality managers within HP perform a variety of functions, from training to communicating the best software engineering and business practices to establishing productivity and quality metrics.

HP has adopted a methodology called Total Quality Control (TQC). A fundamental principle of TQC is that all company activities

| Objectives | Measures | Target | Initiative |
|---|---|---|---|
| Financial | | | |
| Reduce administrative costs | Cost-to-spend ratio | Reduce administrative costs by 10% | Implement plan to improve efficiencies |
| Increase profits | Net earnings | Increase revenue | Improve volume and collection efforts |
| Internal business processes | | | |
| Decrease the number of fraudulent transactions | The number of fraud instances reported | Reduce the number by 10% | Implement stronger policies to penalize dishonest sellers |
| Reduce system downtime | Downtime, in minutes per month | Achieve 99% online per month | Schedule system maintenance during low customer use times |
| Learning and growth | | | |
| Improve internet marketing | Number of hits and site visits from internet ads | Increase by 15% | Assess target market for effectiveness of current ad placements |
| Increase number of buyers and sellers by regions | Number of auctions and bids per auction | Increase by 5% per region | Form region-specific auctions |
| Increase participants in B2C auctions | Number of business selling surplus via auctions | Increase by 10% | Market push to businesses |
| Customer relations | | | |
| Improve response time to respond to questions | Time from receipt of question to response | Decrease to four hours for 99% of responses | Empower first-level CSR's to respond to questions |
| Decrease the number of disputes | Number of resolved and unresolved disputes | Decrease unresolved disputes by 5% | Implement plan to intervene earlier and faster in disputes |

**Figure 7.5**   Typical balanced scorecard measures, targets, and initiatives.

can be scrutinized in terms of the processes involved; metrics can be assigned to each process to evaluate effectiveness. HP established the Systems Software Certification program to ensure measurable, consistent, high-quality software through defining metrics, setting goals, collecting and analyzing data, and certifying products for release.

HP's results were impressive. Defects were caught and corrected early, and costs to find and fix defects were lower. Less time was spent in the costly system test and integration phases, and on maintenance. This resulted in lower overall support costs and higher productivity. It also increased quality for HP's customers.

HP's success demonstrates what a corporate-wide commitment to continuous improvement can achieve. The commitment to these gains was so strong that HP invested in full-time productivity and quality managers, which is indeed unique.

Techniques to introduce quality programs vary from company to company, but there are some commonalities:

1. Do a customer satisfaction survey.
2. Get management sponsorship to fix what you found wrong in the customer satisfaction survey.
3. Top management needs to make a visible and personal commitment to any quality program.
4. Customers as well as suppliers need to be involved.
5. Define what the processes are.
6. Come up with ways to improve the process.
7. Determine metrics to measure the improvement of the process.

### Quality Control

There are several techniques commonly used by businesses to control the quality of products. A check sheet is one of the easiest ways to collect data that can be used in later analysis when determining quality issues. Graphs and histograms are graphical techniques that can be used to organize, summarize, and display data over time.

The old adage that 80 percent of the problems are caused by 20 percent of the causes is the reason for the popularity of the Pareto analysis. This technique visually represents which factor or factors are causing the problem, as shown in Figure 7.6. In this example, "bad data" seems to be the culprit.

To create a Pareto diagram:

1. Select a problem you want to analyze.
2. Determine the categories of the problem and collect the data you want to display.
3. Note the categories on the horizontal axis in descending order of value.
4. Determine measurement scale (cost, frequency, etc.) and note it on the left vertical axis.
5. Optionally draw a cumulative line from left to right that shows the cumulative percent of the categories.

The Ishikawa diagram (also called a fishbone diagram or cause and effect diagram) is a diagram that shows the causes of a certain event.

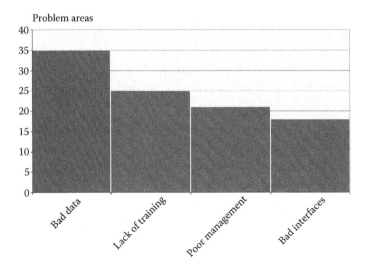

**Figure 7.6**   Sample Pareto analysis.

It was first used in the 1960s and is considered one of the seven basic tools of quality management, along with the histogram, Pareto chart, check sheet, control chart, flowchart, and scatter diagram. Figure 7.7 shows a first pass in creating a fishbone diagram for the problem of late projects.

To create a fishbone diagram:

1. Define the problem clearly and objectively.
2. Write a problem statement in a box at the right of diagram.

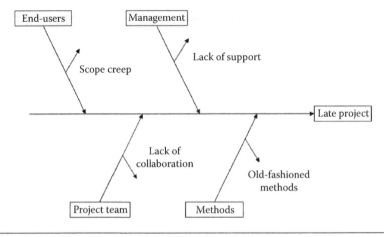

**Figure 7.7**   Fishbone, or cause and effect, diagram.

3. Define the major categories of possible causes (use generic branches). Factors to consider include data and information systems, dollars, environment, hardware, materials, measurements, methods, people, training, and equipment.
4. Construct the diagram by inserting the major categories at the ends of lines.
5. Brainstorm possible and specific causes and list them under the appropriate category.
6. Vote to identify the likely root causes.

### Conclusion

Quite a high percentage of IT projects continue to overrun their schedules and budgets, are canceled, or are not completed on time. Unfortunately, these are not shocking statistics to those who manage IT departments. Although there are many reasons why systems go over budget or over schedule (e.g., changing requirements, departing employees, reallocated functionality), the biggest contributing factor to the problem is simply poor project management and failure to assess and mitigate development and implementation risks. Effective project management includes the processes of project tracking, promoting quality assurance, process improvement, performance measurement and management, strict budget management, and risk planning.

# 8

# FUNDAMENTALS OF INFORMATION TECHNOLOGY PROJECT MANAGEMENT

The Standish Group published its landmark "Chaos Report" in 1995. The report included statistics such as "84% of projects fail or are significantly challenged" and "45% of developed features are never used," and the report is among the most oft-quoted in the industry. A decade later there seems to be some improvement. Most of the troublesome projects in the 2004 Chaos Report had a cost overrun of under 20 percent of the budget, a threefold improvement over the first 1994 study. Of all projects with cost overruns, including failed projects, the average project cost overrun in 2004 was found to be 43 percent, versus an average cost overrun of 180 percent in 1994. Of the projects studies, 53 percent were deemed challenged, 18 percent failed, and the rest were successful. In 2015, the Chaos Report found that 29 percent of projects were successful, 19 percent failed, and 52 percent were challenged. Hence, the risks of information technology (IT) project development continue to remain high.

## Why Project Planning

Although project planning will never eliminate all over-budget and over-schedule situations, the systematic methodology that project planners should utilize will certainly reduce the likelihood of problems and risk.

Like all other aspects of systems development, the development of a project plan (i.e., project planning) cannot be done in a vacuum. A wide variety of people (i.e., stakeholders) need to be involved for the plan to be accurate and workable. Once the project plan is developed, it is used to embark on the systemization of the ideas presented in the plan (i.e., managing the project). The plan itself is never cast in concrete. What this means is that a project plan is often modified, as constraints, assumptions, and even risks change during the life cycle of the systems development effort.

### Project Management and the Systems Development Life Cycle

To understand information technology project management, one must first understand where project planning fits within the traditional systems development life cycle (SDLC). As shown in Figure 8.1, the SDLC is a set of phased processes that see a software development effort from its inception through its implementation.

Projects are similar to living entities: they are born, they live, and then they die. This is why we call it a "life cycle." A system starts out as someone's idea, that is, a "concept." For example, someone in finance might have an idea to build an accounts payable system that processes payments through the Internet. If the idea is deemed feasible, it is placed in development (i.e., systems development and design). Once the system has been designed, it can be coded and then implemented (i.e., placed in production for end-users to use). Eventually, however, systems outlive their usefulness. At this point, they are either retired or replaced. We can then say that the system is "closed out."

Thus, the four generic stages of the project life cycle can be said to be:

1. Concept (i.e., feasibility, project planning)
2. Development (i.e., analysis, design, code, test)
3. Implementation (i.e., conversion, maintenance)
4. Closeout

This is the macro view of the project life cycle. As is often the case, there are variations on this theme, as shown in Table 8.1.

*Concept*

The "idea" phase of the SDLC is the point at which the end-users, systems analyst, and various managers meet for the first time, although the systems analyst might not actually be involved at this point. This is where the scope and objectives of the system are fleshed out in a very high level document.

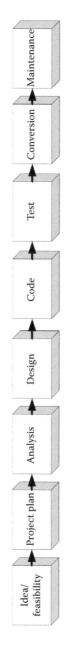

**Figure 8.1**    How the project plan fits within the systems development life cycle (SDLC).

**Table 8.1**   Project Life Cycle

| DISCOVERY | INITIATION | STUDY | DESIGN, BUILD, AND TEST | OPERATE |
|---|---|---|---|---|
| Project value and prioritization | Project approach (build/buy, phases) | Project charter | High-level design | Business build |
| Feasibility study | Request for information | Project kickoff | Low-level design | Human resources (HR) build |
| Estimation | Business case | Business requirements | Product build | Transition to operations |
| IT budget | RFP (request for proposal) for study | System requirements | User procedures | Postimplementation review |
| IT roadmap revision | Source selection | Update business case | Training | Customer satisfaction survey |
| Demand assessed | Contract for study | RFP for build | Acceptance testing | Project closure |
| | Cost/benefits identified | Source selection | Finished product | Support |
| | | Contract for build | | Maintenance requests |
| | | Requirements scoped/cost-benefits updated | | Steady requests |

Next, a team composed of one or more system analysts and end-users try to determine whether the system is feasible. There are many reasons why systems are not feasible: too expensive, technology not yet available, and not enough experience to create the system are just some of the reasons why a system will not be undertaken.

Many metrics are used to determine feasibility. One of the most popular is return on investment (ROI). However, be forewarned; some in the IT field find ROI an inexact science.

Once the system is determined to be feasible, a project plan is completed that details the project's scope, costs, schedule, and resource requirements.

*Development*

Systems analysis can now be initiated using software engineering methodologies. This is the point when the analysts determine the rules and regulations of the system, for example:

- What are the inputs?
- What are the outputs?
- What kind of online screens will there be?
- What kind of reports should there be?
- Will paper forms be required?
- Will any hookups to external files or companies be required?
- How shall this information be processed?

In general, methodologies can be categorized as follows. It should be noted that a methodology can be used in conjunction with another methodology.

- Waterfall—This is a phased, structured approach to systems development. The phases include requirements feasibility, analysis, system design, coding, testing, implementation, and testing. Please note that there are variations of these stated phases. Usually, each phase is performed sequentially although there is some potential for overlap. This is the methodology that is used most often in industry.
- Iterative (prototyping)—Most often this approach is used to replace several of the phases in the waterfall approach. In the

waterfall approach, the time to market, so to speak, can be months (sometimes years). During this time, requirements (scope) may change and the final deliverable, therefore, might be quite outmoded. To prevent this from happening it is a good idea to try to compress the development cycle to shorten this time to market and provide interim results to the end-user. The iterative model consists of three steps: (1) listen to customer, (2) build/revise a mock-up, (3) customer test-drives the mock-up, and then return to step 1.

- Rapid application development (RAD)—This is a form of the iterative model. The key word here is *rapid*. Development teams try to get a first pass of the system out to the end user within 60 to 90 days. To accomplish this, the normal seven-step waterfall is compressed into the following steps: business modeling, data modeling, process modeling, application generation and testing, and turnover. Note the term *application generation*. RAD makes use of application generators, formerly called computer-assisted software engineering (CASE) tools.

- Incremental model—The four main phases of software development are analysis, design, coding, and testing. If we break a business problem into chunks (or increments), then we can use an overlapping, phased approach to software development. For example, we can start the analysis of increment one in January, increment two in June, and increment three in September. Just when increment three starts up, we are at the testing stage of increment one and coding stage of increment two.

- Joint application development (JAD)—JAD is more of a technique than a complete methodology. It can be utilized as part of any of the other methodologies discussed here. The technique consists of one or more end-users who are then "folded" into the software development team. Instead of an adversarial software developer–end-user dynamic, the effect is to have the continued, uninterrupted attention of the person(s) who will ultimately be using the system.

- Reverse engineering—This technique is used to first understand a system from its code, second generate documentation base on the code, and then make desired changes to the system. Competitive software companies often try to reverse engineer their competitors' software.

- Reengineering—Business goals change over time. Software must change to be consistent with these goals. Instead of building a system from scratch, the goal of reengineering is to retrofit an existing system to new business functionality.
- Object-oriented—Object-oriented analysis (OOA), object-oriented design (OOD), and object-oriented programming (OOP) are very different from what we have already discussed. In fact, you will need to learn a whole new vocabulary as well as new diagramming techniques.
- Agile—Agile software development is a methodology in which requirements and solutions evolve through collaboration between self-organizing, cross-functional teams. Agile promotes adaptive planning, evolutionary development, early delivery, and continuous improvement, and it encourages rapid and flexible response to change. Scrum is an iterative and incremental agile software development framework for managing product development. A key principle of Scrum is the understanding that customers will change their minds about what they want or need, and that there will be unpredictable challenges.

*Implementation*

Upon delivery of a systems specification to programmers, implementation can get underway. The systems analyst, project leader, and project manager are all responsible for making sure that the implementation effort goes smoothly. Programmers code code and then test that code. When this first level (unit testing) of testing is done, there are several other phases of testing including systems testing (putting all of the programs in the system together to see how they work as a group), parallel testing (testing the old system versus the new system), and integration testing (testing program-to-program interfaces).

Once the system has been fully tested it is turned over to production (changeover). Usually, just prior to this, the end-user departments (not just the team working on the project) are trained and manuals distributed. The entire team is usually on-call during the first few weeks of the system after changeover since errors often crop up and it can take several weeks for the system to stabilize.

If the development is not targeted for in-house consumption, that is, it is meant to be sold to external customers, another setup is needed for the testing. In many cases, trusty customers of the developing company are approached with the offer to become the testers of the beta version (a version released for testing before its final wrap-up).

Once the system is stabilized it is evaluated for correctness. At this point a list of things to correct as well as a wish list of things that didn't wind up in the first phase of the system is created and prioritized. The team, which consisted of technical and end-user staff, usually stays put and works on future versions of the system. This phase of the SDLC is referred to as maintenance.

*Closeout*

Eventually, all systems reach the end of their utility. They must be either retired or replaced. There are many reasons for closeout, for example:

- Requirements have changed
- Regulations have changed
- New technologies are introduced
- Technologies in use are deemed obsolete
- Functionality has been outsourced
- Functionality has been incorporated into another system

**The Project Planning Document**

The project planning document articulates what the system will do. Some developers are troubled by the fact that the project planning process is usually undertaken prior to the stages of systems analysis and design, where end-user requirements are typically captured. Where the systems analysis and design stages capture a micro view of the system, the project planning process is only required to capture a macro view of the system. This is usually sufficient for the purposes of planning, resource allocation, and budgeting.

The project plan:

- Defines what will be done—The goals and objectives section of the project plan provides a general statement of the scope of the project as well as a more detailed list of requirements, interfaces (e.g., EDI, database), and constraints (e.g., the system must have a response time of sub-one second).

- Clearly defines when it will be done—Most organizations permit only the most experienced of personnel to tackle the difficult task of scheduling. These pros utilize a variety of methodologies (e.g., deterministic approach, stochastic approach) and tool sets (e.g., Microsoft Project) to apportion tasks to available personnel in the most optimal manner, as shown in Figure 8.2.

- Clearly defines how much it will cost—The project plan is bottom-line oriented and, therefore, must include the estimated costs of the project for review and approval by the various stakeholders.

  Estimators may use a wide variety of techniques to perform the project cost effort. It is customary, in fact, for the estimator to use at least two techniques and then "triangulate" the two (i.e., discuss the reasons why the two estimates have differences). Often, the "true" estimate is the average of the results of the various estimation methods used.

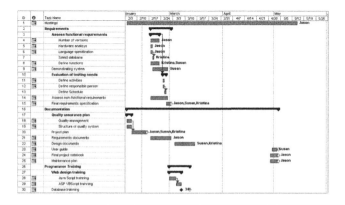

**Figure 8.2** A typical schedule. This one was created using Microsoft Project.

One of the more popular estimation methodologies is COCOMO (Cost Construction Model II). Figure 8.3 shows its application to cost estimation in project planning.

The "true" cost of a project must also include software and hardware required; time expended by outside personnel such as trainers, end-users, and managers; and administrative overhead.

- Clearly defines what resources (people and other) will be needed—The reader of the project plan should be able to easily ascertain:
  - How much the project will cost
  - How long it will take
  - How many people are needed
  - What these people will do
  - What kinds of people (e.g., programmers, trainers) are needed
  - What, if any, software will be required to be purchased
  - What, if any, hardware will be required to be purchased
  - What, if any, outside services will be required to be secured
  - What business, project and/or product risks might be encountered, and the resources required to counter these risks

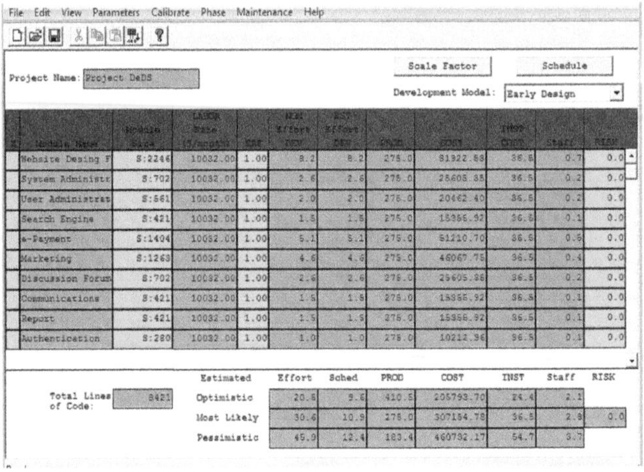

**Figure 8.3**  Using COCOMO to estimate a project's cost.

Roles in Project Management

The people involved in the creation, implementation, and subsequent running of a computer system are called stakeholders. Stakeholders are interrelated. The very best way to get a bird's-eye view of how your stakeholders are organized is to request or build an organization chart for the company. This will provide you with a veritable who's who in your organization.

From a project management perspective, three stakeholder groups are relevant:

1. The project manager
2. Management
3. The project team

*The Project Manager*

There are several categories of IT managers. Starting from the top you have the chief information officer (CIO). This person is usually a senior officer of the company and might report directly to the president or chairman. Sometimes the CIO is called the chief technical officer (CTO). Some organizations have both a CIO and a CTO.

Reporting to the CIO is a series of project managers (PMs). The PM is usually responsible for one or more systems. Reporting to the PM may be one or more project leaders (PLs).

A project leader is responsible for a specific project. Reporting to the project leader is one or more systems analysts and programmers (see Figure 8.4). The project might also call for the services of a graphic designer, web designer, database administrator, and quality assurance person. Usually these additional people do not actually work for the project leader but are merely assigned to the project leader for a specific purpose according to a project plan.

The project manager's responsibilities include:

- Interfaces with other stakeholder groups, including end-users and senior management
- Interfaces with technology staff including project leaders, systems analysts, and programmers
- Develops or aids in developing the project plan

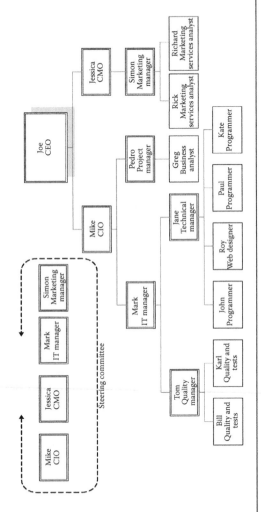

**Figure 8.4**   A typical IT organization and staff breakdown.

- Oversees the feasibility study
- Manages one or more projects
- Maintains the schedule (i.e., allocates and reallocates resources as necessary)
- Keeps the project on budget
- Oversees project tracking and control

*Management*

There are several layers of management that project managers need to be concerned with:

1. Senior managers—The chief executive officer (CEO), president, senior vice presidents, and chief financial officer (CFO) are usually only peripherally involved in a particular project. However, without their initial support the project will never be funded. Without their continued support, the project will never be successfully implemented.
2. Business managers—These are the line managers who oversee the departments that the end-users work for. These managers are instrumental in getting the project proposal elevated and prioritized. In addition, these managers often act as "champions," cajoling reluctant end-users into cooperating with the project team and reluctant senior management into funding the project.
3. IT managers—IT managers work with both senior managers and business managers to determine project priorities. In addition to this responsibility, IT managers (a) oversee all project development efforts, (b) champion new technologies and methodologies, securing funding for them from senior management, and (c) act as a liaison to corporate senior managers as well as business managers.
4. Risk managers—Some organizations invest in risk managers, who are charged with overseeing risk planning in all departments. This manager might not be a member of the IT department.

Without management support and involvement, a project will surely fail.

*The Project Team*

There are a wide variety of people other than managers involved in the process of systems development. The highest level goal of the project team is to successfully implement the system as specified in the project plan. The following discusses the makeup of a typical project team.

- The systems analyst—Today's system analysts need to know both the end-users' business as well as be technically proficient (e.g., know accounting and Java or human resources and C++). It is the systems analyst who is responsible for working with the end-users to determine system feasibility and then develop the scope, requirements, design, and other documents. He or she is then responsible for implementing, testing, and then turning over a completed, working system.
- Programmer—The programmer is usually the most technical person on the systems team. It is this person who knows all of the buzzwords that you see in the "help wanted" ads: Java, C++, Oracle, Sybase, and so on. The role of the programmer is to take the specifications handed to him or her by the systems analyst and turn these specifications into working programs, and ultimately complete systems.
- End-users—The end-user department is composed of experts who do a particular task. Maybe they are accountants or maybe they are in marketing; they still are experts in what they do. A single system may have many end-users who hail from many different departments. Some end-users might not work for the company at all. Each end-user will have a different set of requirements. It is the role of the end-user to work with the systems analyst to uncover and then document these requirements. It is also the job of the end-user to assist in other phases of the SDLC, such as the testing component of the implementation stage.
- Other roles—There are other people involved in the systems development effort. Systems organizations have many departments. Systems analysts and programmers generally work for the application development department. Other departments reporting to the CIO might include the web development department (home to graphic designers and people who can

program using web-based and mobile technologies); quality assurance department (home to people responsible for rigorously testing your system); technical writing department (home to people who might write your policies and procedures guides and other manuals); database administration (home to the database administrator [DBA]); networking department (home to the folks who administer your network); and help desk (home to the people who support your end-users).

Other roles include clients and partners of your company. Many companies partner with organizations and, thus, share data. Electronic Data Interchange (EDI) is an example of this. For example, a jewelry manufacturer will use EDI to process orders from major clients, such as department stores, who purchase its jewelry.

### Project Management Office

As we've learned, project management is a set of discrete steps that sees a project from inception to closure, as shown in Figure 8.5. However, project management should not be performed in a vacuum. A particular project is just one of many projects that will be

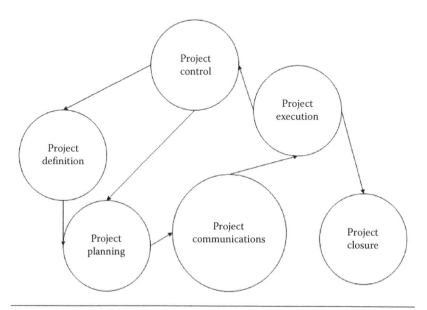

**Figure 8.5**  Project management perspectives.

implemented at any given time. A particular project might be one out of many projects for a specific program. A program is related to a corporate strategy, for example, become an e-book publisher. In our e-book example, there might be multiple projects related to this goal. One project might be to develop a website where e-books could be sold. Another project might be to develop the software that converts print books into e-books.

Most organizations will have several ongoing programs in play all at once—all related to one or more business strategies. It is conceivable that hundreds of projects are ongoing, all in various stages of execution. Portfolio management is needed to provide the business and technical stewardship of all of these programs and their projects, as shown in Figure 8.6.

Portfolio management is often performed by a project management office (PMO). This is the department or group that defines and maintains the standards of process, generally related to project management, within the organization. The PMO strives to standardize and introduce economies of repetition in the execution of projects. The

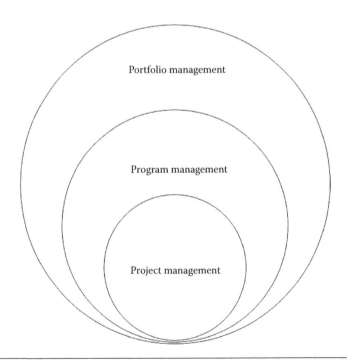

**Figure 8.6**   Portfolio management.

PMO is the source of documentation, guidance, and metrics on the practice of project management and execution.

A good PMO will base project management principles on accepted, industry standard methodologies such as PMBOK or PRINCE2, which are discussed in following sections. Increasingly influential industry certification programs such as ISO 9000 and the Malcolm Baldrige National Quality Award (MBNQA), government regulatory requirements such as Sarbanes-Oxley; and business process management techniques such as the balanced scorecard have propelled organizations to standardize processes.

### Balanced Scorecard

Over the past decade many CIOs have realized that it is not sufficient to manage merely the IT end of the business. The integration of IT strategy to business strategy must be managed as well. As mentioned in the last section, one tool chosen for this task is the balanced scorecard, as shown in Figure 8.7.

As discussed in earlier chapters, Robert S. Kaplan and David P. Norton developed the balanced scorecard approach in the early 1990s to compensate for their perceived shortcomings of using only financial

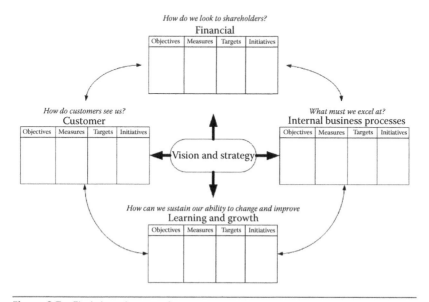

**Figure 8.7** The balanced scorecard.

metrics to judge corporate performance. They recognized that in this new economy it was also necessary to value intangible assets. Because of this they urged companies to measure such esoteric factors as quality and customer satisfaction. By the mid-1990s, the balanced scorecard became the hallmark of a well-run company. Kaplan and Norton often compare their approach for managing a company to that of pilots viewing assorted instrument panels in an airplane cockpit—both have a need to monitor multiple aspects of their working environment.

In the scorecard scenario, a company organizes its business goals into discrete, all-encompassing perspectives: financial, customer, internal process, and learning/growth. The company then determines cause-and-effect relationships, for example, satisfied customers buy more goods, which increases revenue. Next, the company lists measures for each goal, pinpoints targets, and identifies projects and other initiatives to help reach those targets.

Departments create scorecards tied to the company's targets, and employees and projects have scorecards tied to their department's targets. This cascading nature provides a line of sight between each individual, what they are working on, the unit they support, and how that impacts the strategy of the whole enterprise.

For IT managers, the balanced scorecard is an invaluable tool that will finally permit IT to link to the business side of the organization using a cause-and-effect approach. Some have likened the balanced scorecard to a new language, which enables IT and business line managers to think together about what IT can do to support business performance. A beneficial side effect of the use of the balanced scorecard is that, when all measures are reported, one can calculate the strength of relations between the various value drivers.

*The Portfolio Perspective*

Managing from an investment perspective—with a continuing focus on value, risk, cost, and benefits—has helped businesses reduce IT costs by up to 30 percent with a 2 to 3 times increase in value. This is often referred to as portfolio management.

A stepwise plan for implementation follows:

1. Take inventory—A complete inventory of all IT initiatives should be developed. Information such as the project's sponsors and champion, stakeholder list, strategic alignment with corporate objectives, estimated costs, and project benefits should be collected.
2. Analyze—Once the inventory is completed and validated, all projects on the list should be analyzed. A steering committee should be formed that has enough insight into the organization's strategic goals and priorities to place IT projects in the overall strategic landscape.
3. Prioritize—The output of the analysis step is a prioritized project list. The order of prioritization is based on criteria that the steering committee selects. This is different for different organizations. Some companies might consider strategic alignment to be the most important, whereas other companies might decide that cost-benefit ratio is the better criterion for prioritization.
4. Manage—Portfolio management is not a one-time event. It is a constant process that must be managed. Projects must be continually evaluated based on changing priorities and market conditions.

**Project Management Methodologies**

There are two major project management methodologies. The Project Management Body of Knowledge (PMBOK), which is most popular in the United States, recognizes five basic process groups typical of almost all projects: initiating, planning, executing, controlling, and monitoring, and closing. Projects in Controlled Environments (PRINCE2), which is the de facto standard for project management in the United Kingdom and is popular in more than 50 other countries, defines a wide variety of subprocesses, but organizes these into eight major processes: starting a project, planning, initiating a project, directing a project, controlling a stage, managing product delivery, managing stage boundaries, and closing a project.

*PMBOK*

PMBOK is an IEEE standard (1490-2003). It provides a methodology for a wide range of project categories, including software, engineering, construction, and automotive. The five basic process groups and nine knowledge areas are typical of most projects, programs, and operations. Processes overlap and interact throughout a project or phase, and are described in terms of inputs (e.g., documents, plans, designs), tools and techniques, and outputs (e.g., documents, products).

The nine knowledge areas are project integration management, project scope management, project time management, project cost management, project quality management, project human resource management, project communications management, project risk management, and project procurement management.

*PRINCE2*

PRINCE is the UK's government standard for IT project management. The latest version, PRINCE2, was designed for all types of management projects. PRINCE provides a structured approach (i.e., method) to project management, within a clearly defined framework. Similarly to PMBOK, PRINCE2 is a process-driven method. It defines forty-five separate subprocesses organized into eight major processes:

Starting Up a Project (SU)
    SU1 Appointing a Project Board Exec and Project Manager
    SU2 Designing a Project Management Team
    SU3 Appointing a Project Management Team
    SU4 Preparing a Project Brief
    SU5 Defining Project Approach
    SU6 Planning an Initiation Stage
Planning (PL)
    PL1 Designing a Plan
    PL2 Defining and Analyzing Products
    PL3 Identifying Activities and Dependencies
    PL4 Estimating

PL5 Scheduling

PL6 Analyzing Risks

PL7 Completing a Plan

Initiating a Project (IP)

IP1 Planning Quality

IP2 Planning a Project

IP3 Refining the Business Case and Risks

IP4 Setting up Project Controls

IP5 Setting up Project Files

IP6 Assembling a Project Initiation Document

Directing a Project (DP)

DP1 Authorizing Initiation

DP2 Authorizing a Project

DP3 Authorizing a Stage or Exception Plan

DP4 Giving Ad Hoc Direction

DP5 Confirming Project Closure

Controlling a Stage (CS)—Projects should be broken into stages and these subprocesses dictate how each individual stage should be controlled.

CS1 Authorizing Work Package

CS2 Assessing Progress

CS3 Capturing Project Issues

CS4 Examining Project Issues

CS5 Reviewing Stage Status

CS6 Reporting Highlights

CS7 Taking Corrective Action

CS8 Escalating Project Issues

CS9 Receiving Completed Work Package

Managing Product Delivery (MP)

MP1 Accepting a Work Package

MP2 Executing a Work Package

MP3 Delivering a Work Package

Managing Stage Boundaries (SB)

SB1 Planning a Stage

SB2 Updating a Project Plan

SB3 Updating a Project Business Case

SB4 Updating the Risk Log
SB5 Reporting Stage End
SB6 Producing an Exception Plan
Closing a Project (CP)
CP1 Decommissioning a Project
CP2 Identifying Follow-on Actions
CP3 Project Evaluation Review

PRINCE is composed of a number of components, including the business case, or justification behind the project, plans, controls, management of risk, quality, configuration management, and change control.

The business case is a crucial element to the PRINCE2 methodology but is surprisingly absent from PMBOK. However, if one adapts a software engineering approach to project management, as recommended, then the omission of the business case within PMBOK itself should present no problem.

## Conclusion

Projects operate in an environment much broader than the project itself. This means that the project manager needs to understand not only the intricacies of the particular project, but the greater organizational context in which its stakeholders exist. Project managers must identify and understand the needs of all the stakeholders (i.e., project team, management, end-users) while delivering a quality, reduced-risk product on time and within budget.

# 9

# PROJECT CRITICAL
# SUCCESS FACTORS

In this chapter, we will examine project critical success factors (CSFs). Topics highlighted include managing people, dealing with politics, and managing for disaster.

### Just What's Critical to Project Success

Quite a few things can go wrong with software development, as shown in Table 9.1. A different spin on this is shown in Table 9.2.

There are a wide variety of management considerations (e.g., project scope, scheduling, risk, tracking, estimation, etc.) that can make or break a project. We usually call these considerations critical success factors. There is a diversity of opinions on what drives project success. The Standish Group's top 10 reasons are:

1. User involvement
2. Executive management support
3. Clear business objectives
4. Optimizing scope
5. Agile process
6. Project manager expertise
7. Financial management
8. Skilled resources
9. Formal methodology
10. Standard tool and infrastructure

In the discussion that follows, all of these critical success factors, except the last two, are from the Standish Group's study of successful and unsuccessful projects. The last two factors—contract negotiation and management, and implementation—are lessons learned from various information technology (IT) projects. Tables 9.3 through 9.5

**Table 9.1**  Classic Project Mistakes

| PEOPLE-RELATED MISTAKES | PROCESS-RELATED MISTAKES | PRODUCT-RELATED MISTAKES | TECHNOLOGY-RELATED MISTAKES |
|---|---|---|---|
| Undermined motivation | Overly optimistic schedules | Requirements gold-plating, i.e., too many product features | Silver-bullet syndrome, i.e., latching onto a new technology or methodology that is unproven for the particular project |
| Weak personnel | Insufficient risk management | Feature creep | Overestimated savings from tools or methods |
| Uncontrolled problem employees | Contractor failure | Developer gold-plating, i.e., developers using technology just for the sake of using that technology | Switching tools in the middle of a project |
| Heroics | Insufficient planning | Push me, pull me negotiation, i.e., constantly changing schedule | Lack of automated source code control |
| Adding people to a late project | Abandonment of planning under pressure | Research-oriented development, i.e., stretching the limits of technology | |
| Noisy, crowed offices | Wasted time before project actually starts, i.e., during the approval and budgeting process | | |
| Friction between developers and customers | Shortchanged upstream activities, e.g., requirements analysis | | |
| Unrealistic expectations | Inadequate design | | |
| Lack of effective project sponsorship | Shortchanged quality assurance | | |
| Lack of stakeholder buy-in | Insufficient management controls | | |
| Lack of user input | Premature or too frequent convergence, i.e., releasing the product too early | | |
| Politics over substance | Omitting necessary tasks from estimates | | |
| Wishful thinking | Planning to catch up later | | |
| | Code-like-hell programming | | |

**Table 9.2**   Success/Failure Factors

FACTORS RELATED TO PROJECT
Size and value
Having a clear boundary
Urgency
Uniqueness of project activities
Density of the project network (in dependencies between activities)
Project life cycle
End-user commitment
Adequate funds/resources
Realistic schedule
Clear goals/objectives

FACTORS RELATED TO THE PROJECT MANAGER/LEADERSHIP
Ability to delegate authority
Ability to trade-off
Ability to coordinate
Perception of his or her role and responsibilities
Effective leadership
Effective conflict resolution
Having relevant past experience
Management of changes
Contract management
Situational management
Competence
Commitment
Trust
Other communication

FACTORS RELATED TO PROJECT TEAM MEMBERS
Technical background
Communication
Troubleshooting
Effective monitoring and feedback
Commitment

FACTORS RELATED TO THE ORGANIZATION
Steering committee
Clear organization/job descriptions
Top management support
Project organization structure
Functional manager's support
Project champion

(*Continued*)

**Table 9.2 (Continued)**   Success/Failure Factors

FACTORS RELATED TO THE ENVIRONMENT
  Competitors
  Political environment
  Economic environment
  Social environment
  Technological environment
  Nature
  Client
  Subcontractors

**Table 9.3**   Success Factors: Sample Assessment

| SUCCESS FACTOR | STANDISH RANK | STANDISH WEIGHT | ASSESSMENT* | SCORE |
|---|---|---|---|---|
| Executive support | 1 | 18 | 3 | 54 |
| User involvement | 2 | 16 | 3 | 48 |
| Experienced project manager | 3 | 14 | 3 | 42 |
| Clear business objectives | 4 | 12 | 2 | 24 |
| Minimized scope | 5 | 10 | 2 | 20 |
| Agile requirements process | 6 | 8 | 2 | 16 |
| Standard infrastructure | 7 | 6 | 3 | 18 |
| Formal methodology | 8 | 6 | 2 | 12 |
| Reliable estimates | 9 | 5 | 2 | 10 |
| Skilled staff | 10 | 5 | 3 | 15 |
| Contract negotiations and management | – | 10 | 3 | 30 |
| Implementation | – | 8 | 2 | 16 |
| TOTAL SCORE | | | | 301 |
| WEIGHTED SCORE | | | | 86.5% |

*In this example, the assessment is 3, 2, or 1 for high, medium, or low, respectively.

take these success factors and propose a method for assigning a green, yellow, or red indicator to this category.

- Executive support—The executive sponsor must have a global view of the project, set the agenda, arrange the funding, articulate the project's overall objectives, be an ardent supporter, be responsive, and, finally, be accountable for the projects success.
- User involvement—Primary users must have good communication skills allowing them to clearly explain business

**Table 9.4**   Project Outlook Stoplight Criteria

| DASHBOARD AREA | GREEN | YELLOW | RED |
|---|---|---|---|
| Scope | • Total cost of all change requests is 50% or less of change request budget, and<br>• All major system components will be implemented as planned | • Total cost of all change requests is 75% or less of change request budget, or<br>• Major system component will be deferred to later phase in order to meet current phase's schedule or budget | • Total cost of all change requests is at least 75% of the change request budget, or<br>• Major system component will not be implemented |
| Schedule | • Schedule variance does not impact completion date for current phase, and<br>• Work plan is updated at least once every two weeks | • Schedule variance delays completion date for current phase but does not impact completion date for later phases or critical path, or<br>• Major deliverable will be late by two weeks or less, or<br>• Work plan has not been updated within last 30 days | • Schedule variance affects critical path, or<br>• Major deliverable will be at least two weeks late, or<br>• Work plan has not been updated for more than 30 days |
| Budget | • Budget variance is less than 5% of total budget and there is project funding flexibility within the agency's control | • Budget variance is less than 10% of total budget and there is project funding flexibility within the agency's control | • There is a budget variance and there is no remaining project funding flexibility, or<br>• Budget variance is at least 10% |
| Success factors | • Weighted score is at least 90% | • Weighted score is at least 80% | • Weighted score is less than 80% |

processes in detail to the IT organization. Primary users should also be trained to follow project management protocols. Finally, users must be realists and aware of the limitations of the projects.

• Experienced project manager—Project managers must possess technology and business knowledge, judgment, negotiation, good communication, and organization. The focus is on softer skills, such as diplomacy and time management.

**Table 9.5**   Project Outcome Stoplight Criteria

| DASHBOARD AREA | GREEN | YELLOW | RED |
|---|---|---|---|
| Scope | • Project satisfies at least 95% of all business objectives, and<br>• All major system components are implemented as planned | • Project satisfied at least 90% of all business objectives, and<br>• No more than one major system component is deferred to later phase | • Project satisfies less than 90% of all business objectives, or<br>• At least one major system component is not implemented |
| Schedule | • Project completion no later than 10% of original schedule duration | • Project completion no later than 20% of original schedule duration | • Project completion later than 20% of original duration |
| Budget | • Budget variance is less than 5% of total budget | • Budget variance is less than 10% of total budget | • Budget variance is at least 10% of total budget |
| Success factors | • Weighted score is at least 90% | • Weighted score is at least 80% | • Weighted score is less than 80% |

- Clear business objectives—The project objectives must be clearly defined and understood throughout the organization. Projects must be regularly measured against these objectives to provide an opportunity for early recognition and correction of problems, justification for resources and funding, and preventive planning on future projects.
- Minimized scope—The scope must be realistic and able to be accomplished within the identified project duration and measured regularly to eliminate scope creep.
- Agile business requirements process—Requirements management is the process of identifying, documenting, communicating, tracking, and managing project requirements, as well as changes to those requirements. Agile requirements process is the ability to do requirements management quickly and without major conflicts. This is an ongoing process and must stay in lockstep with the development process.
- Standard infrastructure—Establish a standard technology infrastructure that includes operational and organizational protocols. This infrastructure must be commonly understood and regularly assessed.

- Formal methodology—Following formal methodology provides a realistic picture of the project and the resource commitment. Certain steps and procedures are reproducible and reusable maximizing project-wide consistency.
- Reliable estimates—Be realistic.
- Skilled staff—Properly identify the required competencies, the required level of experience and expertise for each identified skill, the number of resources needed within the given skill, and when these will be needed. Soft skills are equally important when identifying competencies.
- Contract negotiation and management—The Standish Group did not identify this as a success factor. However, based on lessons learned from various projects, contract negotiation and management plays a major role in project outcomes.
- Implementation—The Standish Group did not identify this as a success factor. However, based on lessons learned from various projects, implementation plays a major role in project outcomes.

**The Right People**

Having the right people on a project team is certainly key to the success of a project. In a large pharmaceutical company the lead designer walked off a very important project. Obviously, that set the team back quite, a bit as no one else had enough experience to do what he did. Even if the IT staff says put, there is still the possibility that a "people" issue will negatively affect the project. For example, a change in senior management might mean that the project you are working on gets canned or moved to a lower priority. A project manager working for America Online Time Warner had just started an important new project when a new president was installed. He did what all new presidents do: He engaged in a little housecleaning. Projects got swept away, and so did some people. When the dust settled, the project manager personally had a whole new set of priorities, as well as a bunch of new end-users to work with. Although the project manager's most important project stayed high on the priority list, unfortunately, some of the end-users did not. The departure of a subject matter expert can have disastrous consequences. Lucky for our intrepid project manager, she was able to replace her "domain expert" with someone equally knowledgeable.

Today's dynamically changing business landscape can also play havoc with projects. Mergers and acquisitions can have the effect of changing the key players and/or adding whole new sets of stakeholders and stakeholder requirements. Going global adds an entirely new dimension to the importance of being able to speak the end-users' language.

Personnel changes and mergers and acquisitions pale beside the one thing that has the most dramatic effect on the success or failure of our projects: corporate politics. Politics is something that we are all familiar with and are definitely affected by. We cannot change it, so we have to live with it. "Being political" is something we might look down upon. Nonetheless, it is something that we all have to learn to do if we are to shepherd our projects through to successful project completion.

Having the right people on your team and being on a team favored by current management are just two critical success factors.

### Technological Issues

I once worked on a project where the technical lead selected software based on the technology he wanted to put on his resume, rather than what the project required. The software was implemented but had to be replaced within a few years due to frequent system failure. An important critical success factor, as you can surmise, is making sure the technology fits the project.

The team must have the wherewithal to develop the software using the tools and technologies selected. The system itself must be fully specified, with all dependencies listed, so that all developers can readily understand what they need to do and how their piece of the pie relates to what everyone else is doing.

Lest the system fail repeatedly, as ours did, it is a good idea to strive for simple is better. In other words, the architecture should be simple and straightforward, where possible.

Companies should use formal software engineering methodologies, with validation at each major milestone (systems analysis, design, etc.). This ensures that what was designed actually makes its way into production. Along this line, it is best to opt for incremental or agile development, where possible. Massively large systems, with hundreds of processes and programmers, are just too large to effectively manage for a single implementation. It is best to divide this massive project into smaller, more manageable projects instead.

### Effective Communications

One of the most important skills a project manager can have is interpersonal skills. He or she must be able to effectively communicate with a wide variety of people across the entire organization. The project manager must be equally at ease when working with the CEO, as he or she is when working with data entry clerical staff.

The project manager must be able to:

- Make the person being spoken to feel at ease
- Understand the language of the end-user
- Understand the business of the end-user
- Interpret what the end-user is saying correctly and completely
- Write effectively and using the proper style
- Be able to make meaningful presentations
- Be articulate
- Be able to mediate disputes
- Understand the politics of the organization

### The Proper Utilization of Standards

There are many methodologies a project manager might employ when doing his or her job. The Software Engineering Institute's Capability Maturity Model (CMM) defines five levels of software process maturity. The lowest level, Initial, is typified by little formalization. The highest level, Optimized, is defined by the use of quality standards such as formal use of a methodology and process measurement. The project manager should strive to utilize the very highest levels of standards of practice such that the Optimized level of CMM can be achieved.

Software engineering (i.e., development) consists of many components, including definitions, documentation, testing, quality assurance, and metrics. Standards bodies (i.e., ANSI, ISO, IEEE) have crafted standards for many of these.

Standards enable software developers to develop quality-oriented, cost-effective, and maintainable software in an efficient, cost-productive manner. The goal of each standard is to provide the software developer with a set of benchmarks, enabling him or her to complete the task and be assured that it meets at least a minimum level of quality. Indeed, the dictionary definition of standard is "an acknowledged measure of comparison for quantitative or qualitative value; a criterion." Thus, standards provide the developer with the criteria necessary to build a system. It is the role of the project manager to ensure that the proper standards are being adhered to.

*Ethics*

One of the very highest standards a project manager can aspire to achieve is a heightened sense of ethics. The newspapers have been filled with stories of the results of a lapse of ethics (e.g., Parmalat, Enron, Arthur Andersen). When dealing with individuals, the organization as a whole, or society at large, the project manager must:

- Be fair
- Be impartial
- Be honest
- Be forthright

*Being Political*

A project must be developed from a position of strength. Since the project manager is the one in charge of the project, the PM must be powerful, know how to get power, or align him- or herself with a powerful sponsor.

What do you do if political gamesmanship is getting in the way of your project's success? Here are some things that you can do to neutralize opposition or attacks on a project:

- Identify persons who are opposed to the project.
- Determine why they feel the project is not to their advantage.

- Meet with anyone who directly attacks your or the project and ask that person what is troubling him or her. Show this person how his or her actions will affect the project and the organization, and then ask for suggestions to get him or her to support the project.
- Place all agreements and progress reports in writing. This provides an audit trail.
- Speak directly and truthfully; never hedge your comments.
- Distribute a memo to stakeholders, including the opposition, to clarify all rumors. Project opponents frequently use the office rumor mill to distribute misinformation about the project.
- Be prepared to defend all actions that you take. Make sure you have a solid rationale for your decisions.

*Legal*

Legal and regulatory issues will also have an affect on whether the system will ultimately be successful. Examples include copyright, cybercrimes, and spam.

*Organizational*

Computer systems can benefit organizations in many ways. However, some changes required by the introduction of a system might be considered disruptive and, thus, undesirable. If, for example, a system will require that the entire company be reorganized, then this system might be deemed infeasible.

Conclusion

My students studied the issues surrounding project failure and came up with their own list of critical success factors:

- Successfully relating to end-users
- Understanding and dealing with organizational politics
- Resolving conflicts
- Motivating internal stakeholders

- Being a visionary who can see the end product right from the start
- Being able to relate to tech gurus
- Successful team building
- Being a subject matter expert (SME)
- Understanding risk analysis
- Willingness to commit "intelligent disobedience"; this is the ability to disagree with stakeholders that might pose threats leading to the derailment of the project

# 10

# LEGAL, PRIVACY, AND SECURITY RISK

Does the following concern you?

- When you buy a Microsoft Kinect, you are bringing into your home, or office, a telescreen that can recognize who is in the room and interpret body language.
- A joint effort by a British university and a Canadian security company will bring to a theater near you the ability to monitor facial expressions.
- Cisco commissioned a survey of 2,600 workers and information technology (IT) professionals in 13 countries. Twenty percent of IT leaders said that their relationship with their employees is dysfunctional, demonstrating a disconnect between IT, employees, and policies.
- A survey of 1,100 mobile workers found that 22 percent of employees had breached their employers' strict smartphone policies when using nonmanaged personal smartphones to access corporate information.
- One in eight malware attacks are via a USB device, according to the security firm Avast Software.
- The U.S. Department of Defense estimates that over 100 foreign intelligence organizations have attempted to break into U.S.-based networks (government, university and businesses). Every year hackers steal enough data to fill the Library of Congress many times over.
- Viruses can come from any connected device, including MP3 players, cameras, fax machines, and even digital picture frames. In 2008, Best Buy found a virus in the Insignia picture frames that it was selling.

- Companies outsourcing data storage (to a cloud) are responsible for any data that has been breached. So, do make sure the cloud provider or data service provider you use is carefully vetted.
- Cybercriminals are getting smarter. They invented poisoned search results, ransomware, rogue antivirus, social networking malware, malicious advertisements, and even built-in instant messaging clients that are used to notify criminals when the mark has logged into his or her online bank account.

The use of computers raises some risk issues around content use, infringement, defamation, attribution, tort liability, privacy, and security. Although most of these relate to public social networking sites such as Facebook and LinkedIn, some of the issues are still relevant to internal networks, particularly if public platforms are integrated into the tool sets.

### Website Legal Issues

#### Defamation/Torts

Wikis, blogs, workspaces, social networks, and so on provide ample opportunity for defamation (i.e., harming the reputation of another by making a false statement to a third person). These resources should be monitored for this as well as for possibility of other tort liability. Examples of this would be intentional infliction of emotional distress, interference with advantageous economic relations, fraud, or misrepresentation.

#### Trademarks

Trademark or service mark notices should be notably displayed wherever the marks appear. If a mark has is registered with the U.S. Patent and Trademark Office (http://www.uspto.gov/), the registered trademark (®) symbol should be displayed; otherwise, the trademark (™) or service mark (SM) symbols should be displayed. Organizations should be vigilant in protecting its trademarks and service marks. They should be equally vigilant that they do not infringe on the marks of others. Content that resides on the organization's servers need to be audited to make sure that no trademark infringement is taking place.

### Copyright

A copyright is a form of protection provided to the authors of "original works of authorship," including literary, dramatic, musical, artistic, and certain other intellectual works such as software, both published and unpublished. The 1976 Copyright Act generally gives the owner of copyright the exclusive right to reproduce the copyrighted work, to prepare derivative works, to distribute copies or audio recordings of the copyrighted work, to perform the copyrighted work publicly, or to display the copyrighted work publicly.

The copyright protects the form of expression rather than the subject matter of the writing. For example, a description of a machine could be copyrighted, but this would only prevent others from copying the description; it would not prevent others from writing a description of their own or from making and using the machine.

It is important that the organization audit data residing in its social networks to make sure that any content, data and/or information is not violating anyone else's copyright. For example, dynamically accessing Google and downloading research results to a social network is not a good idea. This is because Google's content is copyrighted to Google, you would need to take care that you are not violating any copyrights.

Using any third party content with permission can result in both criminal and civil liability, including treble damages and attorney fees under the U.S. Copyright Act. Essentially, the best tact to take is to periodically review all content, screening for possible copyright violations.

### Computer Fraud and Abuse Act

Most organizations have provided their employees with PCs capable of wireless Internet access. Many companies and home users have installed wireless Internet connectivity in their offices and homes. It is not unusual for people to seek out unsecured "hot spots," as these wireless connections have come to be known. Several computer equipment manufacturers have even developed inexpensive, small hot-spot locaters for this purpose. The Computer Fraud and Abuse Act (CFAA) makes punishable whoever intentionally accesses a computer without authorization. Organizations will have to develop a very clear policy warning employees against using corporate-supplied PCs in this manner.

*Corporate Content*

Not that long ago a Congressman made a secret trip to Iraq. When he got there, he tweeted that he had just landed. Secret no more. Although we have not focused on the use of Twitter as a social enterprising tool, we expect this tool, or Yammer, its corporate equivalent, to be used. Since these sorts of tools enable almost instantaneous communication with an entire network of people, external as well as internal, users need to take care on what exactly they are communicating.

## Developing Your ePolicy

It is important that the organization develop an ePolicy that addresses how employees use e-mail, the Internet, and all things social networking. The ePolicy should be comprehensive and included as part of the employee handbook. This should be reviewed with each new employee. It would also be a good idea to refresh everyone's memory on a yearly basis by sending out an e-mail instructing employees to review the ePolicy. The ePolicy should be stored on the corporate Intranet as well. It is recommended that one person be assigned as the main point of contact for the ePolicy, should any questions or problems arise.

Some of the points that should be addressed in the ePolicy include:

- Whether employees may use the Internet for personal use.
- Whether external social networking services such as Facebook, LinkedIn, or Yammer may be used.
- Information on whether e-mail is being monitored (it should be). Let employees know that e-mail and any social networking system used is owned by the organization and it can be expected that management or others might access e-mail, workspaces, blogs, Wikis, and so forth.

- Specifics about the type of content that can be maintained within any social networking site (external or internal) (e.g., copyrighted materials).
- Netiquette policies for e-mail and use of social networking websites.
- Specifics on corporate discrimination and sexual harassment policies, particularly as it pertains to online environments.
- The fact that individual employees are expected to respect the privacy of the individuals whose information they have access to, and to use all available security methods to preserve the integrity and privacy of information within their control.
- A directive that specifies that employees are not to engage in any activity that alters or damages data, software, or other technological-related resources belonging to the organization or to someone else, compromising another individual's ability to use technological-related resources, or intentionally disrupting or damaging corporate technological-related resources.
- A stipulation that individuals are expected to report potential abuse that they might have observed for appropriate resolution.

**Security Issues**

Not too long ago Cisco commissioned a study on security in the workplace. Its findings are probably not all that surprising to you:

1. One out of five employees altered security settings on work devices to bypass IT policy so they could access unauthorized websites. More than half said they simply wanted to access the site, and one-third said "it's no one's business" which sites they access.
2. Seven out of ten IT professionals said employee access of unauthorized applications and websites ultimately resulted in as many as half of their companies' data loss incidents. This belief was most common in the United States (74%) and India (79%).
3. Two out of five IT pros dealt with employees accessing unauthorized parts of a network or facility. Of those who reported

this issue, two-thirds encountered multiple incidents in the past year and 14 percent encountered this issue monthly.

4. One out of four employees admitted verbally sharing sensitive information with nonemployees, such as friends, family, or even strangers. When asked why, some of the most common answers included, "I needed to bounce an idea off someone," "I needed to vent," and "I did not see anything wrong with it."

5. Almost half of the employees surveyed share work devices with others, such as nonemployees, without supervision.

6. Almost two out of three employees admitted using work computers daily for personal use. Activities included music downloads, shopping, banking, blogging, and participating in chat groups. Half of the employees use personal e-mail to reach customers and colleagues, but only 40 percent said this is authorized by IT.

7. At least one in three employees leave computers logged on and unlocked when they are away from their desk. These employees also tend to leave laptops on their desks overnight, sometimes without logging off, creating potential theft incidents and access to corporate and personal data.

8. One in five employees store system log-ins and passwords on their computer or write them down and leave them on their desk, in unlocked cabinets, or pasted on their computers.

9. Almost one in four employees carry corporate data on portable storage devices outside of the office.

10. More than one in five employees allow nonemployees to roam around offices unsupervised. The study average was 13 percent, and 18 percent have allowed unknown individuals to tailgate behind employees into corporate facilities.

As you can see, information systems are vulnerable to many threats that can inflict various types of damage, resulting in significant losses. This damage can range from errors harming database integrity to fires destroying entire system centers.

Problems can stem from inside the company (wayward employees) to the more common scenario of those outsiders who would do the company harm. All manner of hardware and software is at risk, including mobile devices. In 2010 we all awoke to the news that iPad users' e-mail addresses and device IDs were exposed. In 2009, security experts identified thirty security flaws in the software and operating systems of smartphones. In 2010, two European university researchers extracted an entire database of text messages from an iPhone, including those that had been deleted, using a corrupt website they controlled. Google's Android operating system averaged 5,768 malware attacks daily over a six-month period, according to CYREN's (www.cyren.com) Security Report for 2013.

Losses from these exploits can stem, for example, from the actions of supposedly trusted employees defrauding a system, from outside hackers, or from careless data entry. Organizations should develop an Information Systems Security Program to implement and maintain the most cost-effective safeguards to protect against deliberate or inadvertent acts, including:

- Unauthorized disclosure of sensitive information or manipulation of data
- Denial of service or decrease in reliability of critical information system (IS) assets
- Unauthorized use of systems resources
- Theft or destruction of systems assets

Best practices for a security checklist encompass the following areas: access control, confidentiality, integrity, availability, nonrepudiation, protection, detection, reaction to incidents, configuration management, vulnerability management, personnel security, physical security, security awareness, and training. All of these should be reviewed upon initiation of any development program to set the parameters for use of that program. The checklist should also be used on a periodic basis to ensure the security of the enterprise platform on an ongoing basis.

The organization should develop an IS security plan to meet the following goals:

- Achieve data integrity levels consistent with the sensitivity of the information processed
- Achieve systems-reliability levels consistent with the sensitivity of the information processed
- Comply with applicable state and federal regulations

- Implement and maintain continuity of operations plans consistent with the criticality of user information processing requirements
- Implement and follow procedures to report and act on IS security incidents

Organizations should conduct periodic security to ensure that

- Sufficient controls and security measures are in place to compensate for any identified risks associated with the program/system and/or its environment.
- The program/system is being operated cost-effectively and complies with applicable laws and regulations.
- Program/systems' information is properly managed.
- The program/system complies with management, financial, IT, accounting, budget, and other appropriate standards.

There are two types of security assessments that must be conducted periodically in computer facilities: risk assessments and security reviews. A risk assessment is a formal, systematic approach to assessing the vulnerability of computer assets, identifying threats, quantifying the potential losses from threat realization, and developing countermeasures to eliminate or reduce the threat or reduce the amount of potential loss. Risk assessments are to be conducted whenever significant modifications are made to the system.

There are three major IT security controls: management controls, operational controls, and technical controls. The term *management controls* is used to address those controls that are deemed to be managerial in nature. *Technical controls* are security controls that should be implemented on systems that transmit, process, and store information. *Operational controls* address security controls that are implemented by people and directly support the technical controls and processing environment.

Management controls are necessary to manage the security program and its associated risks. They are nontechnical techniques, driven by policy and process, and are put in place to meet IT protection requirements. Program security policies and system-specific policies are developed to protect sensitive information transmitted, stored, and processed within system components. Program security policies are broad and are developed to establish the security program and enforce security at the program management level. System-specific security policies are detailed and are developed to enforce security at the system level. The information, applications, systems, networks, and resources must be protected from loss, misuse, and unauthorized modification, access, or compromise. All organizations that process, store, or transmit information must develop, implement, and maintain an IT security program to ensure the protection of the information. The program security policy establishes the security program, assigns the appropriate personnel, and outlines the security duties and responsibilities for all individuals in the program.

Operational controls focus on controls implemented and executed by people to improve the security of a particular system. Media controls address the storage, retrieval, and disposal of sensitive materials that should be protected from unauthorized disclosure, modification, or destruction. Media protection is composed of two security requirements: computer output controls and electronic media controls. Computer output controls apply to all printout copies of sensitive information and state that all printout copies of sensitive information should be clearly marked. Electronic media controls should encompass all the controls of printout materials; however, procedures need to be established to ensure that data cannot be accessed without authorization and authentication from electronic media that contain sensitive information.

All personnel with responsibilities for the management, maintenance, operations, or use of system resources and access to sensitive information should have the appropriate management approval. Organizations should have personnel security procedures to specify responsibilities of the security personnel and system users involved in management, use, and operation of the system. The IT staff must be alert and trained in offensive and defensive methods to protect the organization's information assets. Adequate staffing and key position backup are essential to running and maintaining a secure environment. Personnel security also includes establishing and maintaining procedures for enforcing personnel controls, including the following:

- Determining appropriate access levels (logically and physically)
- Ensuring separation of duties (logically and physically) to not compromise system data or thwart technical controls

- Conducting security training and providing awareness tools for all staff
- Issuing and revoking user identifications (IDs) and passwords

Technical controls focus on security controls that the computer system executes. These controls depend on the proper configuration and functionality of the system. The implementation of technical controls, however, always requires significant operational considerations. These controls should be consistent with the management of security within the organization.

When updating the security plan, the organization should refer to the security issues and questions in Table 10.1 to help ensure its plan is current.

**Table 10.1**   Internet Security Issues Checklist

| SECURITY ISSUES/INFORMATION TO BE ADDRESSED |
| --- |

1. Describe the functions (data transfer, forms-based data entry, or browser-based interactive applications, etc.) you are using the Internet to perform.
2. Describe your application categories and how they are integrated with your production systems (e.g., information access = hypertext, multimedia, soft content and data; collaboration = newsgroups, shared documents, and videoconferencing; transaction processing = Internet commerce and links to IT applications).
3. What communication protocols are in use? FTP, HTTP, telnet, or a combination?
4. How do you control access, identification and authorization (I&A), sensitive or private information, no repudiation, and data integrity?
5. Are firewalls and/or proxy servers present? If so, describe the software used.
6. Is data encryption used? Is it hardware or software based?
7. What application languages are being used (HTML, XML, JavaScript, etc.)? Are these static, semidynamic, or dynamic?
8. What database connectivity or application programming interfaces (APIs) are in place?
9. Do you have separate web servers? Describe the hardware and software.
10. Describe what controls are in effect for shared resources, including any of the following: password protection, user groups, smartcards, biometrics, data encryption, callback systems, virus scanners, vulnerability scanners, and intelligent agents.
11. Are user logons/passwords challenged frequently and under a multilevel protection scheme? Do you allow synchronization of passwords for a single sign-on?
12. Are passwords changed on a regular basis? How often? Is this system-controlled or manual?
13. How many people have administrative rights to the application, telecommunications, and web servers? Are these rights separated by function or can a single person access all of these?
14. Are backups performed of Internet application files and data files? How often?
15. Is a contingency plan in place? Has it been tested? How often is it updated?

Web Server Security

Securing the operating system (OS) that the web server runs on is the initial step in providing security for the web server. The web server software only differs in functionality from other applications that reside on a computer. However, since the web server may provide public access to the computer as well as organization wide access, it should be securely configured to prevent the web server and the host computer from being compromised by intruders.

One of the precautions to take when configuring a web server is to never run the web service as a root or administrative user (super user). Web services or applications should never be located at the root of a directory structure but in a component-specific subdirectory to provide optimum access management. The web service should be run with the permissions of a normal user. This would prevent the escalation of privilege if the web server were ever compromised. Also, the file system of the web server (directories and files) should not be configured to have write access for any users other than those internal users that require such access. Other precautions and secure configuration issues to consider when configuring a public web server are as follows:

- The web server should be on a separate local area network with a firewall configuration or demilitarized zone (DMZ) from other production systems.
- The web server should never have a trust relationship with any other server that is not also an Internet-facing server or server on the same local network.
- The web server should be treated as an untrusted host.
- The web server should be dedicated to providing web services only.
- Compilers should not be installed on the web server.
- All services not required by the web server should be disabled.
- The latest vendor software should be used for the web server, including all the latest hot fixes and patches.

The web browser is usually a commercial client application that is used to display information requested from a web server. There should be a standard browser that has been approved for use within the system environment. Because of the security holes in scripting languages, such as JavaScript and ActiveX (Microsoft), it is recommended that all scripting languages not required for official systems operation be disabled within the web browsers.

Network security addresses requirements for protecting sensitive data from unauthorized disclosure, modification, and deletion. Requirements include protecting critical network services and resources from unauthorized use and security-relevant denial of service conditions.

Firewalls provide greater security by enforcing access control rules before connections are made. These systems can be configured to control access to or from the protected networks and are most often used to shield access from the Internet. A firewall can be a router, a personal computer, or a host appliance that provides additional access control to the site. The following firewall requirements should be implemented:

- Firewalls that are accessible from the Internet are configured to detect intrusion attempts and issue an alert when an attack or attempt to bypass system security occurs.
- Firewalls are configured to maintain audit records of all security-relevant events. The audit logs are archived and maintained in accordance with applicable records retention requirements and security directives.
- Firewall software is kept current with the installation of all security-related updates, fixes, or modifications as soon as they are tested and approved.
- Firewalls should be configured under the "default deny" concept. This means that for a service or port to be activated it must be approved specifically for use. By default, the use of any service or communications port without specific approval is denied.
- Only the minimum set of firewall services necessary for business operations is enabled and only with the approval of IT management.
- All unused firewall ports and services are disabled.
- All publicly accessible servers are located in the firewall DMZ or in an area specifically configured to isolate these servers from the rest of the infrastructure.

- Firewalls filter incoming packets on the basis of Internet addresses to ensure that any packets with an internal source address, received from an external connection, are rejected.
- Firewalls are located in controlled access areas.

Routers and switches provide communication services that are essential to the correct and secure transmission of data on local and wide area networks. The compromise of a router or switch can result in denial of service to the network and exposure of sensitive data that can lead to attacks against other networks from a particular location. The following best practice solutions should be applied to all routers and switches throughout an application environment:

- Access to routers and switches is password-protected in accordance with policy guidance.
- Only the minimum set of router and switch services necessary for business operations is enabled and only with the approval of IT management.
- All unused switch or router ports are disabled.
- Routers and switches are configured to maintain audit records of all security-relevant events.
- Router and switch software is kept current by installing all security-related updates, fixes, or modifications as soon as they are tested and approved for installation.
- Any dial-up connection through routers must be made in a way that is approved by the IT management.

All systems should use antivirus (AV) utilities or programs to detect and remove viruses or other malicious code. The AV software must be kept current with the latest available virus signature files installed. AV programs should be installed on workstations to detect and remove viruses in incoming and outgoing e-mail messages and attachments, as well as actively scanning downloaded files from the Internet. Workstation and server disk drives should be routinely scanned for viruses. The following specific restrictions should be implemented to reduce the threat of viruses on systems:

- Traffic destined to inappropriate websites should not be allowed.
- Only authorized software should be introduced on systems.

- All media should be scanned for viruses before introduction to the system. This includes software and data from other activities and programs downloaded from the Internet.
- Original software should not be issued to users but should be copied for use in copyright agreements. At least one copy of the original software should be stored according to configuration management controls.

Table 10.2 provides an outline of topics of a systems security plan. As the security plan is being developed, ask the following reflective questions:

1. Does the plan address logical and physical security of the system?
2. Does the logical security include password protection, data encryption (if applicable), and access profiles to preclude access to the data by unauthorized personnel?

**Table 10.2**   Systems Security Plan Outline

| CONTENTS OF THE SYSTEMS SECURITY PLAN | |
|---|---|
| Outline of topics | 1. Scope—Describe the site, giving location, configuration, operations, and processing supported, and identification of IT units and applications covered by the plan |
| | 2. Definitions—Explain any terms that might not be familiar to all readers |
| | 3. Overall security assessment—Discuss policies and practices, addressing assignment of security responsibilities, personnel security clearance policies, audit reports, and training; also assess current and planned activities for the next year |
| | 4. Site plan and equipment schematic |
| | 5. Sensitive application systems (obtain the following information for each system): date of last system evaluation, date of last system certification or recertification, date of next evaluation or recertification |
| | 6. Summary of the risk analysis reports |
| | 7. Continuity plan(s) |
| | 8. Summary of the security reviews for all types of processing platforms in use |
| | 9. Training needs with action schedule |
| | 10. Other supporting documents (terminal security rules, local security procedures, user handbooks, etc.) |
| Policies and procedures | 1. Physical security of resources |
| | 2. Equipment security to protect equipment from theft and unauthorized use |
| | 3. Software and data security |
| | 4. Telecommunications security |
| | 5. Personnel security |
| | 6. Continuity plans to meet critical processing needs in the event of short-or long-term interruption of service |
| | 7. Emergency preparedness |
| | 8. Designation of an IT security officer/manager |

3. Does the logical security provide for supervisory intervention if needed (determined case by case)?
4. Are negotiable documents or authorizations securely stored?
5. Does the physical security address not only the security of the physical devices but also the building security?
6. Does the physical security address safety and environment issues?
7. Does the security plan address data and application backup procedures?
8. Does the security plan include recovery procedures?
9. Does the security plan include disaster preparedness and recovery procedures? (These may be in a separate plan.)
10. If a department or organization-wide security plan exists, is there a clear delineation of where the system security plan leaves off and the organization plan takes over or vice versa?
11. Does the logical security include separation of duties between functions to prevent potential fraud situations?

**Protecting Mobile Devices**

Many people seem to ignore security policies pertaining to their smartphones. They seem not to realize how they could be exposing themselves, their companies, and their companies' stakeholders to harm. Although mobile devices cannot be totally secured, there are some measures that can be taken to afford a measure of security:

1. Don't use hotel wireless networks to access sensitive information.
2. Hotel wired networks are often wide open to eavesdropping. All packets for a set of rooms, a floor or several floors, or even the whole hotel, can be seen by all other systems on the network. Unprotected packets are prime targets for capture, analysis, and data extraction. It is best to invest in wireless broadband for employees who must travel and bring their work with them.
3. Encrypt all data on a device in case it is stolen or lost, seemingly a common occurrence. Better yet, do not store any information at all on the device. Store it on the server or in the cloud.

4. Configure devices to block external snooping. Firewalls are a must. Firewalls are also available for many handheld devices.

5. Back up critical information. It sounds obvious, but those on the road might neglect to do this. If the organization does not have its own mobile accessible backup server, then use a cloud service such as Microsoft OneDrive (https://onedrive.live.com/).

6. Do not start a laptop with a USB device attached. This can result in malware loaded directly to the computer ahead of some antivirus software.

7. Secure all wireless access points. Strong, mixed passwords should be used and changed on a frequent basis.

### Conclusion

Security, legal, and privacy policies need to be created and rigorously enforced. In this chapter, we provided some tools and techniques to do so. Lax security, failure to adhere to legal regulations, and privacy violations open the organization to risk that can be readily avoided.

# 11

# ASSESSMENT AND MITIGATION OF RISKS IN A BRING YOUR OWN DEVICE (BYOD) ENVIRONMENT

Shadow IT (information technology) occurs when end-users buy hardware and create software on their own without the knowledge of the IT department. A 2015 Cisco report shows that corporate IT leaders at large enterprises estimated that their companies used an average of 51 cloud services, when the actual number turned out to be an astonishing 730.

There are several factors that drive shadow IT: understaffed IT departments; a difficult and often bureaucratic approvals process; end-users who know a thing about developing code and/or easy to find and inexpensive developers (e.g., programmers obtained through services like guru.com are often located in India); and BYOD (bring your own device).

When end-users do their own thing, risk is introduced into the organization. These risks include substandard development techniques, an overreliance on cloud provider security, unsecured shadow file storage, and the use of prehacked shadow (unauthorized) USB drives.

As technology becomes increasingly sophisticated so do the risks. The Internet of Things (IoT) is a case in point. Shop floor operations in manufacturing companies often feature programmable robotics for more efficient assembly operations. These machines are usually connected to company networks as well, making them every bit as susceptible to malicious attacks as other network nodes.

Traveler's Insurance (https://www.travelers.com/business-insurance /technology), which provides technology insurance that covers several risk areas such as bodily injury risk, technology errors and omissions risk, and cyber risk, provides an example of the bodily risk scenario:

> A municipal employee uses real-time technical support via a browser-based chat window on his personal PC, which he brings to work. His PC isn't a member of the network, so his security protection is significantly lower than his work computer. A hacker intercepts his chat window, impersonates tech support and encourages him to divulge sensitive credentials (social engineering). The hacker then uses the captured credentials to breach the city's traffic control system, causing accidents and injuries.

BYOD solutions include not just the device and the applications (apps) on that device. Mobile devices (e.g., laptops, phones, tablets) are also used to connect to organizational systems. This chapter illustrates how a hypothetical organization deals with computer security issues in its enterprise operating environment.

### A Process for Controlling Risk in a Bring Your Own Device (BYOD) Environment

In the real world, many solutions exist for computer security problems. No single solution can solve similar security problems in all environments. Likewise, the solutions presented in this example may not be appropriate for all environments. Essentially, there are two methodologies that are used to ensure that mobile devices are securely accessing the network. Agent-based solutions and network-based mobile device management (MDM). In the network-based MDM paradigm, no agents are actually stored on the client device. Instead, network devices are intelligent enough to make security decisions based on user identify, device type, location, and time. Both should be supported. This example can be used to help understand how security issues are examined, how some potential solutions are analyzed, how their cost and benefits are weighed, and ultimately how management accepts responsibility for risks.

This case study about a fictitious company we will call "A Typical Organization" is provided for illustrative purposes only and should not be construed as guidance or specific recommendations to solving specific security issues. Because a comprehensive example attempting to illustrate all security topics would be inordinately long, this example necessarily simplifies the issues presented and omits many details. For instance, to highlight the similarities and differences among controls in the different processing environments, it addresses some of the major types of processing platforms linked together in a distributed system: personal computers, local-area networks (LANs), wide-area networks (WANs), and mainframes; it does not show how to secure these platforms.

This chapter also highlights the importance of management's acceptance of a particular level of risk; this will, of course, vary from organization to organization. It is management's prerogative to decide what level of risk is appropriate, given operating and budget environments and other applicable factors.

*Initiating the Risk Assessment*

A Typical Organization has information systems that comprise and are intertwined with several different kinds of assets valuable enough to merit protection. System components owned and operated by A Typical Organization are assets, as are personnel information, contracting and procurement documents, draft regulations, internal correspondence, and a variety of other day-to-day business documents, memos, databases, and reports. A Typical Organization's assets include intangible elements as well, such as reputation of the organization and the confidence of its employees that information will be handled properly.

A recent change in the directorship of A Typical Organization has brought in a new management team. Among the new chief information officer's first actions was appointing a computer security program manager who immediately initiated a comprehensive risk analysis to assess the soundness of A Typical Organization's computer security program in protecting the organization's assets and its compliance with organizational and regulatory directives. This analysis drew upon prior risk assessments, threat studies, and applicable internal control reports. The computer security program manager also established a timetable for periodic reassessments.

Since the WAN and mainframe used by A Typical Organization are owned and operated by other organizations, they were not treated in the risk assessment as A Typical Organization's assets. And although A Typical Organization's personnel, buildings, and facilities are essential assets, the computer security program manager considered them to be outside the scope of the risk analysis.

After examining A Typical Organization's computer system, the risk assessment team identified specific threats to A Typical Organization's assets, reviewed A Typical Organization's safeguards against those threats, identified the vulnerabilities of those policies, and recommended specific actions for mitigating the remaining risks to A Typical Organization's computer security. The following sections provide highlights from the risk assessment. The assessment addressed many other issues at the programmatic and system levels.

*A Typical Organization's Computer System*

A Typical Organization relies on the distributed computer systems and networks shown in Figure 11.1. They consist of a collection of components, some of which are systems in their own right. Some belong to A Typical Organization, but others are owned and operated by other organizations. This following describes these components,

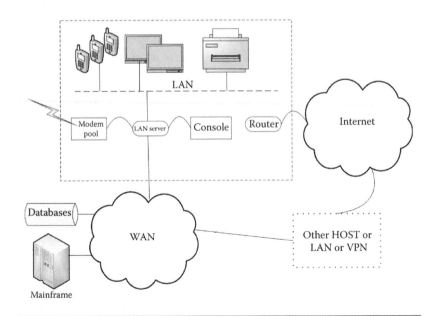

**Figure 11.1**  Typical system architecture.

their role in the overall distributed system architecture, and how they are used by A Typical Organization.

*System Architecture*   A Typical Organization's staff (a mix of clerical, technical, and managerial staff) are provided with personal computers (PCs) located in their offices. Some employees use their own devices out in the field or at home, which typically consists of tablets, laptops, and smartphones.

The devices are connected to a local area network (LAN) so that users can exchange and share information. The central component of the LAN is a LAN server, a more powerful computer that acts as an intermediary between devices on the network and provides a large volume of disk storage for shared information, including shared application programs. The server provides logical access controls on potentially sharable information via elementary access control lists. These access controls can be used to limit user access to various files and programs stored on the server. Some programs stored on the server can be retrieved via the LAN and executed on a PC or other device; others can only be executed on the server. To initiate a session on the network or execute programs on the server, users must log into the server and provide a user identifier and password known to the server. Then they may use files to which they have access.

Since A Typical Organization lets employees access information via BYOD devices, the LAN also provides a connection to the Internet via a router. The router is a network interface device that translates between the protocols and addresses associated with the LAN and the Internet. The router also performs network packet filtering, a form of network access control.

*System Operational Authority/Ownership*   The system components contained within the large dashed rectangle shown in Figure 11.1 are managed and operated by an organization within A Typical Organization known as the Computer Operations Group (COG).

Devices connect to the organization for word processing, data manipulation, and other common applications, including spreadsheet and project management tools. Many of these tasks are concerned with data that are sensitive with respect to confidentiality or integrity.

Some of these documents and data also need to be available in a timely manner.

*Threats to a Typical Organization's Assets*

Different assets of A Typical Organization are subject to different kinds of threats. Some threats are considered less likely than others, and the potential impact of different threats may vary greatly. The likelihood of threats is generally difficult to accurately estimate. Both A Typical Organization and the risk assessment's authors have attempted to the extent possible to base these estimates on historical data but have also tried to anticipate new trends stimulated by emerging technologies.

A Typical Organization's building facilities and physical plant are several decades old and are frequently under repair or renovation. As a result, power; air-conditioning; and LAN, VPN, or WAN connectivity for the server are typically interrupted several times a year for periods of up to one work day. For example, on several occasions, construction workers have inadvertently severed power or network cables. Fires, floods, storms, and other natural disasters can also interrupt computer operations, as can equipment malfunctions.

Another threat of small likelihood but significant potential impact is that of a malicious or disgruntled employee or outsider seeking to disrupt time-critical processing (e.g., payroll) by deleting necessary inputs or system accounts, misconfiguring access controls, planting computer viruses, or stealing or sabotaging computers or related equipment. Such interruptions, depending upon when they occur, can prevent time and attendance data from getting processed and transferred to the mainframe before the payroll processing deadline.

Other kinds of threats may be stimulated by the growing market for information about an organization's employees or internal activities. Individuals who have legitimate work-related reasons for access to the master employee database may attempt to disclose such information to other employees or contractors or to sell it to private investigators, employment recruiters, the press, or other organizations. A Typical Organization considers such threats to be moderately likely and of low to high potential impact, depending on the type of information involved.

Many of the human threats of concern to A Typical Organization originate from insiders. Nevertheless, A Typical Organization also recognizes the need to protect its assets from outsiders. Such attacks may serve many different purposes and pose a broad spectrum of risks, including unauthorized disclosure or modification of information, unauthorized use of services and assets, or unauthorized denial of services.

As shown in Figure 11.1, A Typical Organization's systems are connected to the three external networks: (1) the Internet, (2) the interorganization WAN, and (3) the public-switched (telephone) network. Although these networks are a source of security risks, connectivity with them is essential to A Typical Organization's mission and to the productivity of its employees; connectivity cannot be terminated simply because of security risks.

In each of the past few years before establishing its current set of network safeguards, A Typical Organization had detected several attempts by outsiders to penetrate its systems. Most, but not all of these, have come from the Internet, and those that succeeded did so by learning or guessing user account passwords. In two cases, the attacker deleted or corrupted significant amounts of data, most of which were later restored from backup files. In most cases, A Typical Organization could detect no ill effects of the attack, but concluded that the attacker may have browsed through some files. A Typical Organization also conceded that its systems did not have audit logging capabilities sufficient to track an attacker's activities. Hence, for most of these attacks, A Typical Organization could not accurately gauge the extent of penetration.

In one case, an attacker made use of a bug in an e-mail utility and succeeded in acquiring system administrator privileges on the server—a significant breach. A Typical Organization found no evidence that the attacker attempted to exploit these privileges before being discovered two days later. When the attack was detected, COG immediately contacted the A Typical Organization's incident handling team, and was told that a bug fix had been distributed by the server vendor several months earlier. To its embarrassment, COG discovered that it had already received the fix, which it then promptly installed. It now believes that no subsequent attacks of the same nature have succeeded.

**Table 11.1**   Threat Assessment

| POTENTIAL THREAT | PROBABILITY | IMPACT |
|---|---|---|
| Accidental loss/release of disclosure-sensitive information | Medium | Low/medium |
| Accidental destruction of information | High | Medium |
| Loss of information due to virus contamination | Medium | Medium |
| Misuse of system resources | Low | Low |
| Theft | High | Medium |
| Unauthorized access to telecom resources | Medium | Medium |
| Natural disaster | Low | High |

Although A Typical Organization has no evidence that it has been significantly harmed to date by attacks via external networks, it believes that these attacks have great potential to inflict damage. A Typical Organization's management considers itself lucky that such attacks have not harmed its reputation and the confidence of the citizens its serves. It also believes the likelihood of such attacks via external networks will increase in the future.

A Typical Organization's systems also are exposed to several other threats that, for reasons of space, cannot be fully enumerated here. Examples of threats and A Typical Organization's assessment of their probabilities and impacts include those listed in Table 11.1.

*Current Security Measures*

A Typical Organization has numerous policies and procedures for protecting its assets against the aforementioned threats. These are articulated in A Typical Organization's Computer Security Manual.

A Typical Organization's Computer Operations Group (COG) is responsible for controlling, administering, and maintaining the computer resources owned and operated by A Typical Organization. These functions are depicted in Figure 11.1 enclosed in the large, dashed rectangle. Only individuals holding the job title "System Administrator" are authorized to establish log-in IDs and passwords on multiuser A Typical Organization systems (e.g., the LAN server). Only A Typical Organization's employees and contract personnel may use the system, and only after receiving written authorization from the department supervisor (or, in the case of contractors, the contracting officer) to whom these individuals report.

COG issues copies of all relevant security policies and procedures to new users. Before activating a system account for new users, COG requires that they (1) attend a security awareness and training course or complete an interactive computer-aided instruction training session and (2) sign an acknowledgment form indicating that they understand their security responsibilities.

Authorized users are assigned a secret log-in ID and password, which they must not share with anyone else. They are expected to comply with all of A Typical Organization's password selection and security procedures (e.g., periodically changing passwords). Users who fail to do so are subject to a range of penalties.

Users creating data that are sensitive with respect to disclosure or modification are expected to make effective use of the automated access control mechanisms available on A Typical Organization computers to reduce the risk of exposure to unauthorized individuals. (Appropriate training and education are in place to help users do this.) In general, access to disclosure-sensitive information is to be granted only to individuals whose jobs require it.

A Typical Organization requires various organizations within it to develop contingency plans, test them annually, and establish appropriate administrative and operational procedures for supporting them. The plans must identify the facilities, equipment, supplies, procedures, and personnel needed to ensure reasonable continuity of operations under a broad range of adverse circumstances.

*Contingency Planning*

The continuity department (CD) is responsible for developing and maintaining a contingency plan that sets forth the procedures and facilities to be used when physical plant failures, natural disasters, or major equipment malfunctions occur sufficient to disrupt the normal use of A Typical Organization's PCs, LAN, server, router, printers, and other associated equipment.

The plan prioritizes applications that rely on these resources, indicating those that should be suspended if available automated functions or capacities are temporarily degraded. CD personnel have identified backup facilities sufficient to support A Typical Organization's system-based operations for a few days during an emergency.

No communication devices or network interfaces may be connected to A Typical Organization's systems without written approval of the CD manager.

To protect against accidental corruption or loss of data, CD personnel back up the LAN server's disks onto magnetic tape every night and transport the tapes weekly to a sister organization for storage. A Typical Organization's policies also stipulate that all BYOD users are responsible for weekly backing up any significant data stored on their devices. For the past several years, CD has issued a yearly memorandum reminding users of this responsibility.

To prevent more limited computer equipment malfunctions from interrupting routine business operations, CD maintains an inventory of spare devices. CD is also responsible for reviewing audit logs generated by the server, identifying audit records indicative of security violations, and reporting such indications to the incident-handling team. The CD manager assigns these duties to specific members of the staff and ensures that they are implemented as intended.

The CD manager is responsible for assessing adverse circumstances and for providing recommendations to A Typical Organization's director. Based on these and other sources of input, the director will determine whether the circumstances are dire enough to merit activating various sets of procedures called for in the contingency plan.

*Divisional Contingency Planning*

A Typical Organization's divisions also must develop and maintain their own contingency plans. The plans must identify critical business functions, the system resources and applications on which they depend, and the maximum acceptable periods of interruption that these functions can tolerate without significant reduction in A Typical Organization's ability to fulfill its mission. The head of each division is responsible for ensuring that the division's contingency plan and associated support activities are adequate.

For each major application used by multiple divisions, a chief of a single division must be designated as the application owner. The designated official (supported by his or her staff) is responsible for addressing that application in the contingency plan and for coordinating with other divisions that use the application.

If a division relies exclusively on computer resources maintained by CD (e.g., the LAN), it need not duplicate CD's contingency plan, but is responsible for reviewing the adequacy of that plan. If CD's plan does not adequately address the division's needs, the division must communicate its concerns to the CD director. In either situation, the division must make known the criticality of its applications to the CD. If the division relies on computer resources or services that are not provided by CD, the division is responsible for (1) developing its own contingency plan or (2) ensuring that the contingency plans of other organizations (e.g., the WAN service provider) provide adequate protection against service disruptions.

The LAN server operating system's access controls provide extensive features for controlling access to files. These include group-oriented controls that allow teams of users to be assigned to named groups by the system administrator. Group members are then allowed access to sensitive files not accessible to nonmembers. Each user can be assigned to several groups according to their need to know. (The reliable functioning of these controls is assumed, perhaps incorrectly, by A Typical Organization.)

All users undergo security awareness training when first provided accounts on the server. Among other things, the training stresses the necessity of protecting passwords. It also instructs users to log off the server before going home at night or before leaving the PC unattended for periods exceeding an hour.

*Protection against Network-Related Threats*

A Typical Organization's current set of external network safeguards has only been in place for a few months. The basic approach is to tightly restrict the kinds of external network interactions that can occur by funneling all traffic to and from external networks through interfaces that filter out unauthorized kinds of interactions.

Figure 11.1 shows that the network router is the only direct interface between the LAN and the Internet. The router is a dedicated special-purpose computer that translates between the protocols and addresses associated with the LAN and the Internet. Internet protocols, unlike those used on the WAN, specify that packets of information coming from or going to the Internet must carry an indicator of the kind of service that is being requested or used to process the information. This makes it possible for the router to distinguish e-mail packets from other kinds of packets, for example, those associated with a remote log-in request. The router has been configured by CD to carefully authorize all remote log-in sessions.

A Typical Organization relies on systems and components that it cannot directly control because they are owned by other organizations. A Typical Organization has developed a policy to avoid undue risk in such situations. The policy states that system components controlled and operated by organizations other than A Typical Organization may not be used to process, store, or transmit A Typical Organization information without obtaining explicit permission from the application owner and the CD manager. Permission to use such system components may not be granted without written commitment from the controlling organization that A Typical Organization's information will be safeguarded commensurate with its value, as designated by A Typical Organization. This policy is somewhat mitigated by the fact that A Typical Organization has developed an issue-specific policy on the use of the Internet, which allows for its use for e-mail with outside organizations and access to other resources (but not for transmission of A Typical Organization's proprietary data).

*Vulnerabilities Reported by the Risk Assessment Team*

The risk assessment team found that many of the risks to which A Typical Organization is exposed stem from (1) the failure of individuals to comply with established policies and procedures or (2) the use of automated mechanisms whose assurance is questionable because of the ways they have been developed, tested, implemented, used, or maintained. The team also identified specific vulnerabilities in A Typical Organization's policies and procedures for protecting against payroll fraud and errors, interruption of operations, disclosure and brokering of confidential information, and unauthorized access to data by outsiders.

When someone enters a password to the server, the password is sent to the server by broadcasting it over the LAN "in the clear." This allows the password to be easily intercepted by any other device connected to the LAN. In fact, so-called password sniffer programs that capture passwords in this way are widely available. Similarly, a malicious program planted on a PC could also intercept passwords before transmitting them to the server. An unauthorized individual who obtained the captured passwords could then run applications. Users might also store passwords in a log-on script file.

According to the risk assessment, the server's access controls, with prior caveats, probably provide acceptable protection against unauthorized modification of data stored on the server. The assessment concluded that a WAN-based attack involving collusion between an employee of A Typical Organization and an employee of the WAN service provider, although unlikely, should not be dismissed entirely, especially since A Typical Organization has only cursory information about the service provider's personnel security practices and no contractual authority over how it operates the WAN.

The access control on the mainframe is strong and provides good protection against intruders breaking into a second application after they have broken into a first. However, previous audits have shown that the difficulties of system administration may present some opportunities for intruders to defeat access controls.

The risk assessment concluded that A Typical Organization's safeguards against accidental corruption and loss of time and attendance data were adequate, but that safeguards for some other kinds of data were not. The assessment included an informal audit of a dozen randomly chosen devices (PCs and other devices) and users in the organization. It concluded that many users store significant data on their devices, but do not back them up. Based on anecdotal evidence, the assessment's authors stated that there appear to have been many past incidents of loss of information stored on devices and predicted that such losses would continue.

A Typical Organization takes a conservative approach toward protecting information about its employees. Since information brokerage is more likely to be a threat to large collections of data, A Typical Organization's risk assessment focused primarily, but not exclusively, on protecting the mainframe.

The risk assessment concluded that significant, avoidable information brokering vulnerabilities were present, particularly due to A Typical Organization's lack of compliance with its own policies and procedures. Documents were typically not stored securely after hours, and few PCs were routinely locked. Worse yet, few were routinely powered down, and many were left logged into the LAN server overnight. These practices make it easy for an A Typical Organization employee wandering the halls after hours to browse or copy information on another employee's desk, PC hard disk, or LAN server directories.

The risk assessment pointed out that information sent to or retrieved from the server is subject to eavesdropping by other PCs and devices on the LAN. The LAN hardware transmits information by broadcasting it to all connection points on the LAN cable. Moreover, information sent to or retrieved from the server is transmitted in the clear, that is, without encryption. Given the widespread availability of LAN sniffer programs, LAN eavesdropping is trivial for a prospective information broker and, hence, is likely to occur.

Last, the assessment noted that A Typical Organization's databases were stored on the mainframe, where they might be a target for information brokering by employees of the organization that owns the mainframe. It might also be a target for information brokering, fraudulent modification, or other illicit acts by any outsider who penetrates the mainframe via another host on the WAN.

The risk assessment concurred with the general approach taken by A Typical Organization, but identified several vulnerabilities. It reiterated previous concerns about the lack of assurance associated with the server's access controls and pointed out that these play a critical role in A Typical Organization's approach. The assessment noted that the e-mail allows a user to include a copy of any otherwise accessible file in an outgoing mail message. If an attacker accessed the server and succeeded in logging in as an A Typical Organization employee, the attacker could use the mail utility to export copies of all the files accessible to that employee. In fact, copies could be mailed to any host on the Internet.

The assessment also noted that the WAN service provider may rely on microwave stations or satellites as relay points, thereby exposing A Typical Organization's information to eavesdropping. Similarly, any information, including passwords and mail messages, transmitted during a dial-in session is subject to eavesdropping.

*Recommendations for Mitigating the Identified Vulnerabilities*

The following discussions were chosen to illustrate a broad sampling of topics. Risk management and security program management themes are integral throughout, with particular emphasis given to the selection of risk-driven safeguards.

To remove the vulnerabilities related to unauthorized access to data, the risk assessment team recommended the use of stronger authentication mechanisms based on smart tokens to generate one-time passwords that cannot be used by an interloper for subsequent sessions. Such mechanisms would make it very difficult for outsiders (e.g., from the Internet) who penetrate systems on the WAN to use them to attack the mainframe. The assessment team also recommended improving the server's administrative procedures and the speed with which security-related bug fixes distributed by the vendor are installed on the server.

Although the immaturity of the LAN server's access controls was judged a significant source of risk, CD was only able to identify one other PC LAN product that would be significantly better in this respect. Unfortunately, this product was considerably less friendly to users and application developers, and incompatible with other applications used by A Typical Organization. The negative impact of changing PC LAN products was judged too high for the potential incremental gain in security benefits. Consequently, A Typical Organization decided to accept the risks accompanying use of the current product, but directed CD to improve its monitoring of the server's access control configuration and its responsiveness to vendor security reports and bug fixes.

A Typical Organization concurred that risks of fraud due to unauthorized modification of data at or in transit to the mainframe should not be accepted unless no practical solutions could be identified. After discussions with the mainframe's owning organization, A Typical Organization concluded that the owning organization was unlikely to adopt the advanced authentication techniques advocated in the risk assessment.

The assessment recommended that CD institute a program of periodic internal training and awareness sessions for CD personnel having contingency plan responsibilities. The assessment urged that CD undertake a rehearsal during the next three months in which selected parts of the plan would be exercised. The rehearsal should include attempting to initiate some aspect of processing activities at one of the designated alternative sites. A Typical Organization's management agreed that additional contingency plan training was needed for CD personnel and committed itself to its first plan rehearsal within three months.

After a short investigation, A Typical Organization's divisions, which owned applications that depend on the WAN, concluded that WAN outages, although inconvenient, would not have a major impact on A Typical Organization. This is because the few time-sensitive applications that required WAN-based communication with the mainframe were originally designed to work with magnetic tape instead of the WAN and could still operate in that mode; hence, courier-delivered magnetic tapes could be used as an alternative input medium in case of a WAN outage. The divisions responsible for contingency planning for these applications agreed to incorporate into their contingency plans both descriptions of these procedures and other improvements.

A Typical Organization's management agreed to improve adherence to its virus-prevention procedures. It agreed (from the point of view of the entire organization) that information stored on external devices is frequently lost. It estimated, however, that the labor hours lost as a result would amount to less than a person year, which A Typical Organization management does not consider to be unacceptable. After reviewing options for reducing this risk, A Typical Organization concluded that it would be cheaper to accept the associated loss than to commit significant resources in an attempt to avoid it.

A Typical Organization concurred with the risk assessment's conclusions about its exposure to information-brokering risks and adopted most of the associated recommendations. The assessment recommended that A Typical Organization improve its security awareness training (e.g., via mandatory refresher courses) and that it institute some form of compliance audits. The training should be sure to stress the penalties for noncompliance.

The assessment recommended that an activity log be installed on the server (and regularly reviewed). Moreover, it would avoid unnecessary reliance on the server's access-control features, which are of uncertain assurance. The assessment noted, however, that this strategy conflicts with the desire to store most information on the server's disks so that it is backed up routinely by CD personnel. (This could be offset by assigning responsibility to someone other than the PC owner to make backup copies.) Since the security habits of A Typical Organization's PC users have generally been poor, the assessment also recommended use of hard-disk encryption utilities to protect disclosure-sensitive information on unattended PCs from browsing by unauthorized individuals. Also, ways to encrypt information on the server's disks would be studied.

The assessment recommended that A Typical Organization require stronger authentication for remote access, which would prevent a user from including files in outgoing mail messages; replace its current modem pool with encrypting modems and provide each dial-in user with such a modem; and work with the mainframe organization to install a similar encryption capability for server-to-mainframe communications over the network.

## Conclusion

As illustrated, effective computer security requires clear direction from upper management. Upper management must assign security responsibilities to organizational elements and individuals, and must formulate or elaborate the security policies that become the foundation for the organization's security program. These policies must be based on an understanding of the organization's mission priorities, and the assets and business operations necessary to fulfill them. They must also be based on a pragmatic assessment of the threats against these assets and operations. A critical element is assessment of threat likelihoods. These are most accurate when derived from historical data but must also anticipate trends stimulated by emerging technologies.

A good security program relies on an integrated, cost-effective collection of physical, procedural, and automated controls. Cost-effectiveness requires targeting these controls at the threats that pose the highest risks while accepting other residual risks. The difficulty of applying controls properly and in a consistent manner over time has been the downfall of many security programs. This chapter has provided numerous examples in which major security vulnerabilities arose from a lack of assurance or compliance. Hence, periodic compliance audits, examinations of the effectiveness of controls, and reassessments of threats are essential to the success of any organization's security program.

## Bibliography

Special Publication 800-12. An Introduction to Computer Security: The NIST Handbook. http://csrc.nist.gov/publications/nistpubs/800-12/800 -12-html/chapter20-printable.html.

# 12
## SOCIAL MEDIA RISK

Social media has been defined in a number of ways. For purposes of this chapter, which is adapted from the Federal Financial Institutions Examination Council (FFIEC, 2013), social media is a form of interactive online communication in which users can generate and share content through text, images, audio, and/or video. Social media can take many forms, including, but not limited to, microblogging sites (e.g., Facebook, Google Plus, MySpace, and Twitter); forums, blogs, customer review websites, and bulletin boards (e.g., Yelp); photo and video sites (e.g., Flickr and YouTube); sites that enable professional networking (e.g., LinkedIn); virtual worlds (e.g., Second Life); and social games (e.g., FarmVille and CityVille). Social media can be distinguished from other online media in that the communication tends to be more interactive.

In 2016–2017 Facebook initiated an investigation into possible foreign involvement in the U.S. election. It found fake groups, fake likes and comments, and automated posting across the network by unnamed malicious actors. Essentially, large numbers of fake accounts have been used to strategically disseminate political propaganda and mislead voters. It is obvious that social media applications can be weaponized with very little skill. Social media warfare has become a burden that both government and corporations need to face.

Organizations may use social media in a variety of ways, including marketing; providing incentives; facilitating applications for new accounts; inviting feedback from the public; and engaging with existing and potential customers, for example, by receiving and responding to complaints, or providing pricing. Since this form of customer interaction tends to be both informal and dynamic, and may occur in a less secure environment, it can present some unique challenges.

The use of social media by an organization to attract and interact with customers can impact an organization's risk profile. The increased risks can include the risk of harm to consumers, compliance and legal risk, operational risk, and reputation risk. Increased risk can arise from a variety of directions, including poor due diligence, oversight, or control on the part of the organization.

### Compliance Risk Management Expectations for Social Media

An organization should have a risk management program that allows it to identify, measure, monitor, and control the risks related to social media. The size and complexity of the risk management program should be commensurate with the breadth of the organization's involvement in this medium. For instance, an organization that relies heavily on social media to attract and acquire new customers should have a more detailed program than one using social media only to a very limited extent. However, in accordance with its own risk assessment, an organization that has chosen not to use social media should still consider the potential for negative comments or complaints that may arise within the many social media platforms described earlier, and, when appropriate, evaluate what, if any, action it will take to monitor for such comments and/or respond to them.

The risk management program should be designed with participation from specialists in compliance, technology, information security, legal, human resources, and marketing. Organizations should also provide training for employees' official use of social media. Components of a risk management program should include the following:

- A governance structure with clear roles and responsibilities, whereby the board of directors or senior management direct how using social media contributes to the strategic goals of the organization (for example, through increasing brand awareness, product advertising, or researching new customer bases) and establish controls and ongoing assessment of risk in social media activities, such as threat actor analysis as shown in Figures 12.1 and 12.2.
- Policies and procedures (either stand-alone or incorporated into other policies and procedures) regarding the use and monitoring of social media and compliance with all applicable consumer protection laws and regulations, as appropriate.

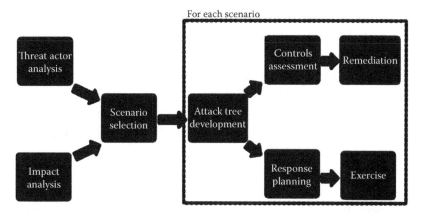

**Figure 12.1**   Threat actor analysis.

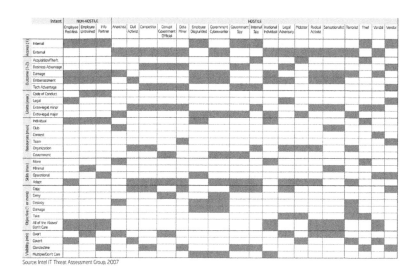

**Figure 12.2**   Intel's threat agent template.

Further, policies and procedures should incorporate methodologies to address risks from online postings, edits, replies, and retention.

- A risk management process for selecting and managing third-party relationships in connection with social media.
- An employee training program that incorporates the organization's policies and procedures for official, work-related use

of social media, and potentially for other uses of social media, including defining impermissible activities.

- An oversight process for monitoring information posted to proprietary social media sites administered by the organization or a contracted third party.
- Audit and compliance functions to ensure ongoing compliance with internal policies and all applicable laws and regulations, as appropriate.
- Parameters for providing appropriate reporting to the organization's board of directors or senior management that enable periodic evaluation of the effectiveness of the social media program and whether the program is achieving its stated objectives.

### Risk Areas

The use of social media to attract and interact with customers can impact an organization's risk profile, including risk of harm to consumers, compliance and legal risks, operational risks, and reputation risks. Increased risk can arise from poor due diligence, oversight, or control on the part of the organization. As previously noted, this chapter is meant to help organizations identify potential risks to ensure organizations are aware of their responsibilities to address risks within their overall risk management program.

#### Compliance and Legal Risks

Compliance and legal risk arise from the potential for violations of, or nonconformance with, laws, rules, regulations, prescribed practices, internal policies and procedures, or ethical standards. These risks also arise in situations in which the organization's policies and procedures governing certain products or activities may not have kept pace with changes in the marketplace. This concern is particularly pertinent to an emerging medium like social media. Further, the potential for defamation or libel risk exists where there is a broad distribution of information exchanges. Failure to adequately address these risks can expose an organization to enforcement actions and/or civil lawsuits.

The laws and regulations discussed in this chapter do not contain exceptions regarding the use of social media. Therefore, to the extent that an organization uses social media to engage in lending, deposit services, or payment activities, it must comply with applicable laws and regulations as when it engages in these activities through other media. Organizations should remain aware of developments involving such laws and regulations.

Section 5 of the Federal Trade Commission (FTC) Act prohibits "unfair or deceptive acts or practices in or affecting commerce." An act or practice can be unfair, deceptive, or abusive despite technical compliance with other laws. An organization should not engage in any advertising or other practice via social media that could be deemed unfair, deceptive, or abusive. Of course, any determination as to whether an act or practice engaged in through social media is unfair, deceptive, or abusive, will necessarily be fact-specific. As with other forms of communication, an organization should ensure that information it communicates on social media sites is accurate, consistent with other information delivered through electronic media, and not misleading.

If social media is used to facilitate a consumer's use of payment systems, an organization should keep in mind the laws, regulations, and industry rules regarding payments that may apply, including those providing disclosure and other rights to consumers. Under existing law, no *additional* disclosure requirements apply simply because social media is involved (for instance, providing a portal through which consumers access their accounts at an organization). Rather, the organization should continue to be aware of the existing laws, regulations, chapter, and industry rules that apply to payment systems, and evaluate those that will apply. These may include the following:

- Electronic Fund Transfer Act/Regulation E—The Electronic Fund Transfer Act (EFTA) and its implementing Regulation E provide specific protections, including required disclosures and error resolution procedures, to individual consumers who engage in "electronic fund transfers" and "remittance transfers."
- Rules Applicable to Check Transactions—When a payment occurs via a check-based transaction rather than an EFT, the

transaction will be governed by applicable industry rules and the Uniform Commercial Code of the relevant state, as well as the Expedited Funds Availability Act, as implemented by Regulation CC (regarding the availability of funds and collection of checks).

Organizations should also be aware of emerging areas of risk in the virtual world. For example, illicit actors are increasingly using Internet games involving virtual economies, allowing gamers to cash out, as a way to launder money. Virtual world Internet games and digital currencies present a higher risk for money laundering and terrorist financing, and should be monitored accordingly.

*Privacy*

Privacy rules have particular relevance to social media when, for instance, an organization collects, or otherwise has access to, information from or about consumers. An organization should take into consideration the following laws and regulations regarding the privacy of consumer information.

Title V of the Gramm-Leach-Bliley Act (GLBA) establishes requirements relating to the privacy and security of consumer information. Whenever an organization collects, or otherwise has access to, information from or about consumers, it should evaluate whether these rules will apply. The rules have particular relevance to social media when, for instance, an organization integrates social media components into customers' online account experience or takes applications via social media portals. An organization using social media should clearly disclose its privacy policies as required under GLBA.

Even when there is no "consumer" or "customer" relationship triggering GLBA requirements, an organization will likely face reputation risk if it appears to be treating any consumer information carelessly or if it appears to be less than transparent regarding the privacy policies that apply on one or more social media sites that the organization uses.

The Controlling the Assault of Non-Solicited Pornography and Marketing Act of 2003 (CAN-SPAM Act) and Telephone Consumer Protection Act (TCPA) may be relevant if an organization sends unsolicited communications to consumers via social media. The CAN-SPAM

Act and TCPA, and their implementing rules, establish requirements for sending unsolicited commercial messages ("spam") and unsolicited communications by telephone or short message service (SMS) text message, respectively. Organizations should be familiar with the provisions of the CAN-SPAM Act and TCPA to evaluate whether social media activities trigger the application of either or both laws.

The Children's Online Privacy Protection Act (COPPA) and the Federal Trade Commission's implementing regulation impose obligations on operators of commercial websites and online services directed to children younger than 13 that collect, use, or disclose personal information from children, as well as on operators of general audience websites or online services with actual knowledge that they are collecting, using, or disclosing personal information from children under 13. An organization should evaluate whether it, through its social media activities, could be covered by COPPA.

Certain social media platforms require users to attest that they are at least 13, and an organization using those sites may consider relying on such policies. However, the organization should still take care to monitor whether it is actually collecting any personal information of a person under 13, such as when a child under 13 manages to post such information on the organization's site.

An organization maintaining its own social media site (such as a virtual world) should be especially careful to establish, post, and follow policies restricting access to the site to users 13 or older, especially when those sites could attract children under 13. This may be true, for instance, in the case of virtual worlds and any other features that resemble video games.

*Reputation Risk*

Recently, a university faculty member shared a snippet of what one student posted in class as a response to an assignment about how to handle a disgruntled customer.

> I would create a review letting everyone know how terrible the apartment was, even if some of it was not true. It would not matter if the owner did everything in their power to make me feel comfortable with the situation. I would try to destroy their business. And that would be by spreading the wrong information and suing.

Reputation risk is the risk arising from negative public opinion and, as we see from the student response, might be false sentiment. Activities that result in dissatisfied consumers and/or negative publicity could harm the reputation and standing of the organization, even if the organization has not violated any law. Privacy and transparency issues, as well as other consumer protection concerns, arise in social media environments. Therefore, an organization engaged in social media activities is expected to be sensitive to, and properly manage, the reputation risks that arise from those activities. Reputation risk can arise in areas including the following.

*Fraud and Brand Identity*   Organizations should be aware that protecting their brand identity in a social media context can be challenging. Risk may arise in many ways, such as through comments made by social media users, spoofs of organization communications, and activities in which fraudsters masquerade as the organization. Organizations should consider the use of social media monitoring tools and techniques to identify heightened risk and respond appropriately. Organizations should have appropriate policies in place to monitor and address in a timely manner the fraudulent use of the organization's brand, such as through phishing or spoofing attacks.

*Third-Party Concerns*   Working with third parties to provide social media services can expose organizations to substantial reputation risk. An organization should regularly monitor the information it places on social media sites. This monitoring is the direct responsibility of the organization, as part of a sound compliance management system, even when such functions may be delegated to third parties. Even if a social media site is owned and maintained by a third party, consumers using the organization's part of that site may blame the organization for problems that occur on that site, such as uses of their personal information they did not expect or changes to policies that are unclear. The organization's ability to control content on a site owned or administered by a third party and to change policies regarding information provided through the site may vary depending on the particular site and the contractual arrangement with the third party. An organization should thus weigh these issues against the benefits of using a third party to conduct social media activities. An organization should conduct an evaluation and perform due diligence appropriate to the risks posed by the prospective service provider prior to engaging with the provider. To understand the risks that may arise from a relationship with a given third party, the organization should be aware of matters such as the third party's reputation in the marketplace; the third party's policies, including policies on collection and handling of consumer information; the process and frequency by which the third party's policies may change; and what, if any, control the organization may have over the third party's policies or actions.

*Privacy Concerns*  Even when an organization complies with applicable privacy laws in its social media activities, it should consider the potential reaction by the public to any use of consumer information via social media. The organization should have procedures to address risks from occurrences such as members of the public posting confidential or sensitive information (for example, account numbers) on the organization's social media page or site.

*Consumer Complaints and Inquiries*  Although an organization can take advantage of the public nature of social media to address customer complaints and questions, reputation risks exist when the organization does not address consumer questions or complaints in a timely or appropriate manner. Further, the participatory nature of social media can expose an organization to reputation risks that may arise when users post critical or inaccurate statements. Compliance risk can also arise when a customer uses social media to communicate issues or concerns directly with an organization.

An organization is expected to take into account the results of its own risk assessments in determining the appropriate approach to take regarding monitoring of, and responding to, such communications. Appropriate steps may include, for example, establishing one or more specific channels consumers must use when submitting complaints or disputes directly to the organization for further investigation, to the extent consistent with other applicable legal requirements. However, the organization should also consider the risks, particularly the reputation risk, inherent in not responding to complaints and disputes received through other channels and tailor its policies and procedures accordingly, in a manner appropriate to the organization's size and risk profile.

Based on its own risk assessment processes, an organization should also consider whether and how to respond to communications disparaging the organization on other parties' social media sites. One approach to managing these risks would be to monitor question and complaint forums on social media sites to ensure that such inquiries, complaints, or comments are reviewed, and when appropriate, addressed in a timely manner.

*Employee Use of Social Media Sites*  Organizations should be aware that employees' communications via social media may be viewed by the public as reflecting the organization's official policies or may otherwise reflect poorly on the organization, depending on the form and content of the communications. Employee communications can also subject the organization to compliance risk and operational risk, as well as reputation risk. Therefore, as appropriate, organizations should take steps to address these risks, such as establishing policies and training to address employee participation in social media representing the organization. For example, if an employee is communicating with a customer regarding a loan product through an approved social media channel, policies should include steps to ensure the customer is receiving all of the required disclosures. This chapter does not address any employment law principles that may be relevant to employee use of social media. In addition, the chapter is not intended to impose any specific requirements for policies or procedures regarding employee personal use of social media. Each organization should evaluate the risks for itself and determine appropriate policies to adopt in light of those risks.

*Operational Risk*

Operational risk is the risk of loss resulting from inadequate or failed processes, people, or systems. The root cause can be either internal or external events. Operational risk includes the risks posed by an organization's use of information technology, which encompasses social media.

Social media is one of several platforms vulnerable to account takeover and the distribution of malware. An organization should ensure that the controls it implements to protect its systems and safeguard customer information from malicious software adequately address social media usage. Organizations' incident response protocol regarding a security event, such as a data breach or account takeover, should include social media, as appropriate.

## Conclusion

As previously noted, this chapter is intended to help organizations understand and successfully manage the risks associated with use of social media. Organizations are using social media as a tool to generate new business and provide a dynamic environment to interact with consumers. As with any product channel, organizations are expected to manage potential risks to the organization and consumers by ensuring that their risk management programs provide appropriate oversight and control to address the risk areas discussed within this chapter.

# Reference

Federal Financial Institutions Examination Council (FFIEC). 2013, January. Social media: Consumer compliance risk management guidance. Docket No. FFIEC-2013-0001. http://www.ffiec.gov/press/Doc/FFIEC%20 social%20media%20guidelines%20FR%20Notice.pdf.

# Index